AF900233

SWAN'S NEST, SWAFFHAM, NORFOLK

Studies in European Archaeology

Series Editors

Joshua C White, David Jacques and Graeme Davis

PETER LANG
Oxford • Bern • Berlin • Bruxelles • New York • Wien

SWAN'S NEST, SWAFFHAM, NORFOLK

EXCAVATING THE PREHISTORY OF THE BRECKLAND CLAYS

Joshua C White

PETER LANG
Oxford • Bern • Berlin • Bruxelles • New York • Wien

Bibliographic information published by Die Deutsche Nationalbibliothek.
Die Deutsche Nationalbibliothek lists this publication in the Deutsche National-bibliografie; detailed bibliographic data is available on the Internet at http://dnb.d-nb.de.

A catalogue record for this book is available from the British Library.

Library of Congress Cataloging-in-Publication Data

Names: White, Joshua C, 1993- author.
Title: Swan's Nest, Swaffham, Norfolk : excavating the prehistory of the Breckland Clays / Joshua C White.
Other titles: Excavating the prehistory of the Breckland Clays
Description: Oxford ; New York : Peter Lang, 2022. | Series: Studies in European archaeology; vol 1 | Includes bibliographical references and index.
Identifiers: LCCN 2021028350 (print) | LCCN 2021028351 (ebook) | ISBN 9781800791046 (hardback) | ISBN 9781800791053 (ebook) | ISBN 9781800791060 (epub)
Subjects: LCSH: Swan's Nest Site (Swaffham, England) | Bronze age--England--Swaffham. | Iron age--England--Swaffham. | Swaffham (England)--Antiquities. | Excavations (Archaeology)--England--Swaffham.
Classification: LCC GN778.22.G7 W53 2022 (print) | LCC GN778.22.G7 (ebook) | DDC 936.26/14--dc23
LC record available at https://lccn.loc.gov/2021028350
LC ebook record available at https://lccn.loc.gov/2021028351

Cover image: Gary Collyer.
Cover design by Brian Melville for Peter Lang Ltd.

ISSN 2754-2440
ISBN 978-1-80079-104-6 (print)
ISBN 978-1-80079-105-3 (ePDF)
ISBN 978-1-80079-106-0 (ePub)

© Peter Lang Group AG 2022
Published by Peter Lang Ltd, International Academic Publishers,
52 St Giles, Oxford, OX1 3LU, United Kingdom
oxford@peterlang.com, www.peterlang.com

Joshua C White has asserted his right under the Copyright, Designs and Patents Act, 1988, to be identified as Author of this Work.

All rights reserved.

All parts of this publication are protected by copyright.
Any utilisation outside the strict limits of the copyright law, without the permission of the publisher, is forbidden and liable to prosecution.
This applies in particular to reproductions, translations, microfilming, and storage and processing in electronic retrieval systems.

This publication has been peer reviewed.

Contents

Figures

Plates

Tables

Contributors

Principal Author	Joshua C White	NPS Archaeology
Graphics	David Dobson	NPS Archaeology
Prehistoric pottery	Sarah Percival	Freelance Specialist
Struck flint	Andrew Peachey	Wardell Armstrong LLP
Non-flint stone	Simon Timberlake	Freelance Specialist
Fired clay	Sue Anderson	Spoilheap Archaeology
Animal bone and molluscs	Julie Curl	Sylvanus
Plant macrofossils	Val Fryer	Freelance Specialist
Pollen, diatoms, charcoal and non-flint stone	Frances Green	Freelance Specialist
Soil micromorphology	Rowena Y Banerjea	QUEST, University of Reading
Aerial imagery analysis	Chris Cox	Air Photo Services

TOM PHILLIPS

Foreword

Archaeologically, later Bronze Age Norfolk offers very contrasting strands of evidence. On the one hand, the widespread distribution of bronze axes, swords, rapiers and other metalwork across large parts of the county, but especially in the west, attests to a highly structured and wealthy society, and presumably a sizeable population. On the other hand, occupation sites and a wider settlement pattern have been much harder to identify, even in the era of developer-funded archaeology. Ashwin (1996: 53–4) acknowledged the wealth of metalwork but declared that later Bronze Age settlement sites in Norfolk were even harder to detect than in preceding periods and could not easily be detected by aerial photography. For a long time, the rich midden deposits at Grimes Graves, with its large accumulations of Deverel-Rimbury pottery and animal bone, remained the most significant evidence of settlement remains excavated in the county. However, in recent years the picture has started to change, helped in part by the discovery of an enclosure system and settlement at Ormesby St Michael in the Norfolk Broads and an accompanying reappraisal of cropmark evidence from elsewhere in the county (Gilmour et al. 2014). Since Ormesby was excavated, a number of other significant Middle-Late Bronze Age settlement sites have been excavated in Norfolk, and one of these sites is Swan's Nest, Swaffham.

This volume presents the results of the open area excavations at Swan's Nest, examining in detail the site's key components – its monumental double-ditched 'Ormesby-type' enclosure, post-built structures and burnt mound evidence – and making the case for the symbolic importance of the enclosure to the community who constructed the original earthworks in the Middle Bronze Age. The prominence of the Swan's Nest enclosure is counterbalanced by the apparent lack of a landscape-wide field system, such as those found further west along the fen-edge environment of the Ouse Valley, around Peterborough and in parts of the Cam Valley. Whether a wider field system exists at Swan's Nest is still to be determined – an appraisal of cropmark evidence in the surrounding landscape is included – but regardless of the bigger picture, what is detailed here makes a significant contribution to the study of later Bronze Age settlements in East Anglia, highlighting the variability that exists not only across the region, but also within Norfolk itself.

Summary

NPS Archaeology undertook a programme of archaeological excavation prior to residential development at Swan's Nest, land west of Watton Road in Swaffham, Norfolk. The site is located at the south-eastern edge of Swaffham, a market town in the west of Norfolk situated c.23.5 km south-east of King's Lynn. The excavation targeted the remains of a large rectangular double-ditched enclosure, first identified by geophysical survey at the site in 2011. In October 2018, following trial trenching at the site, a c.4.2 ha area was opened to expose the remains of the enclosure monument for excavation.

The pre-enclosure activity at the site consisted of a transient Mesolithic presence followed by small-scale Early Neolithic occupation, with a 'developed' pit cluster identified. Investigations on the enclosure established that it was created and in use during the Middle Bronze Age, with activity appearing to reduce after approximately two centuries. Environmental evidence suggests that the Middle Bronze Age enclosure occupied an open landscape of rough grasses, with little apparent arable cultivation occurring in the vicinity. It is considered most likely that a pastoral impetus lies behind the enclosure's creation, though its monumental form, scale and elaborate nature suggests it probably also had an important symbolic function to the community to which it belonged. Several structures were identified within the enclosure, along with significant 'burnt mound' activity, possibly associated with large-scale cooking, industry or ritual practices.

The enclosure remained open into the first millennium BC, with sporadic use of the site occurring in the Late Bronze Age and a more significant reestablishment of activity taking place during the earlier Iron Age. Activity during the earlier Iron Age included a slight modification to the south-west of the enclosure, but beyond this, the nature of this phase was not precisely definable from the excavated evidence. Evidence for anthropogenic use of the site after the earlier Iron Age was limited, with it being unlikely that the site underwent significant agrarian exploitation until the eighteenth/early nineteenth century.

The Swan's Nest enclosure has significant potential to contribute towards the interpretation of the 'Ormesby-type' enclosure monument class; a task of regional importance in forwarding our collective understanding of Bronze Age societies and the establishment of settled agricultural communities in East Anglia. These remains also represent the most substantial discovery of prehistoric activity from the Swaffham area to date, and consequently will be integral in shaping our understanding on the nature of the communities that once occupied the Breckland landscape throughout prehistory.

Acknowledgements

The author would like to thank Abel Homes Ltd for commissioning and funding this archaeological project. Thanks are also given to Steve Hickling of Norfolk County Council Community and Environment Services for monitoring the fieldwork and post-excavation elements of this project.

Thanks are given to Professor Graeme Davis and Professor David Jacques of The University of Buckingham, Tom Phillips of Oxford Archaeology, Lucy Melville of Peter Lang, Dr Zoe Outram of Historic England and Sophie Tremlett of Norfolk County Council Community and Environment Services for their advice and support throughout this project. Frank Davies of Durham University is thanked for the pollen preparations.

The fieldwork was carried out by Richard Williams, Rhys Morgan, Gary Collyer, Scott Crane, Julie Walker, Mick Boyle and the author. The team's perseverance through a harsh winter and the skill with which they excavated the remains is gratefully acknowledged by the author. This publication stands as a testament to their abilities and their dedication to the profession. The author would also like to thank the specialists that contributed towards this project; Sue Anderson, Dr Rowena Y Banerjea, Chris Cox, Julie Curl, David Dobson, Val Fryer, Dr Frances Green, Andrew Peachey, Sarah Percival, Rebecca Sillwood and Dr Simon Timberlake. Thanks are also extended to the author's colleagues at NPS Archaeology; Daniel Fordham, Karen O'Hara, David Adams, David Whitmore, Lilly Hodges, Mick Boyle, David Dobson, Gary Collyer and Amy Duckworth, whose continued support made this publication possible. Special thanks are given to Gary Collyer who produced the reconstruction illustration of the site. Digitising of the drawn records was carried out by David Whitmore, Jenn Hulse and the author. David Adams managed this project on behalf of NPS Archaeology. Professor David Jacques, Steve Hickling, Lilly Hodges and David Whitmore are thanked for their comments on the final manuscript.

CHAPTER 1

Introduction

Archaeological investigations at Swan's Nest, Swaffham, in the county of Norfolk, revealed prehistoric activity dating from the Mesolithic to earlier Iron Age. The most significant remains dated to the Middle Bronze Age and consisted of a large 'Ormesby-type' double-ditch enclosure of regional significance. This publication details the results of the investigations, discusses the significance of these findings within a broader regional context and assesses the activity recorded in relation to the development of the local landscape across time. New information is presented on the 'Ormesby-type' enclosure monument class, a Middle Bronze Age phenomenon specific to the eastern region (after Gilmour et al. 2014), and a new understanding of the later prehistory of the modern locality of Swaffham is provided.

The archaeological excavation of Swan's Nest was prompted by Norfolk County Council Community and Environment Services, following the submission of a proposal for residential development at the site. This development forms part of the incremental southern expansion of Swaffham to have taken place over recent years, which has provided an opportunity through archaeological investigation to construct a picture of the past communities that once occupied this part of the Breckland landscape. Evidence from Swan's Nest has been crucial to this unravelling, with the site appearing to have potentially acted as a central point in the landscape for much of later prehistory. Although the post-prehistoric activity at the site is arguably negligible, its scale and character have been important in determining the full landscape history of the site and its environs, and consequently it has been summarised and presented in this volume.

Topography and Setting

The site is located in the north of the Brecks – a unique landscape region traditionally characterized by open steppe-like heathland stretching across Norfolk and Suffolk, from Swaffham to Bury St Edmunds (Breckland Society 2021). Centred on National Grid Reference TF 8247 0755, the site is situated on the southern edge of Swaffham, a small market town in the west of Norfolk, *c.*23.5 km south-east of King's Lynn (Figure 1). It is bordered to the east by Watton Road which runs in a north-west-to-south-east direction. Immediately to the south and west of the site are housing developments (both still under construction at the time of the excavation). Residential properties are also present immediately north at Swan's Nest House.

The site extended across two agricultural fields; the one to the north measured *c.*3.1 ha in size and the one to the south measured *c.*4.5 ha. The excavation area, which measured a total of *c.*4.2 ha, was spread across these two fields with an east-to-west oriented hedgerow broadly dividing it into two halves (with Area A located to its north and Area B to its south (see Figure 3)). A large oval-shaped pond approximately 24 m × 11 m was present in the excavation area, just south-west of the centre of Area A. This part of the site was not included within the excavation as it was to be incorporated into the subsequent development.

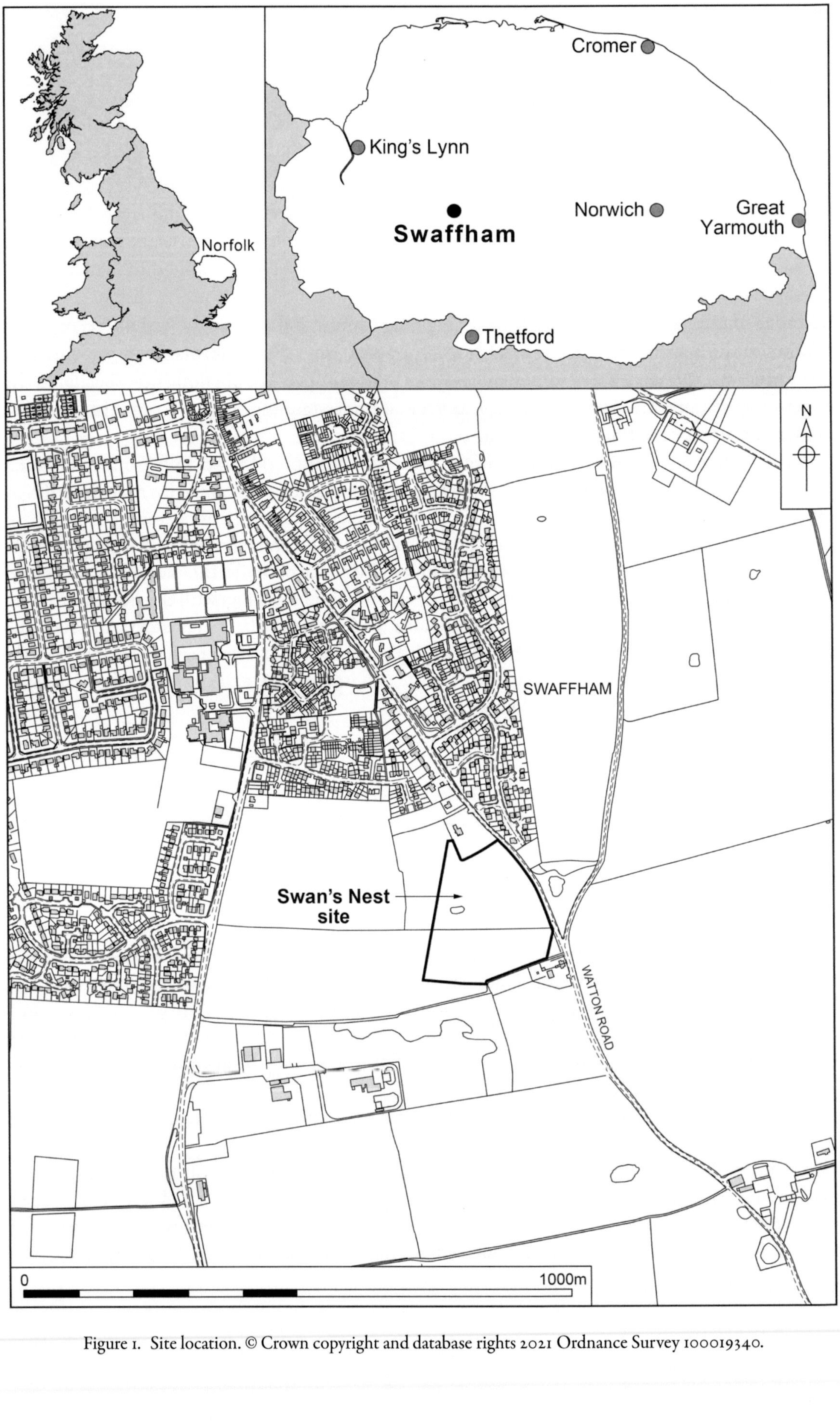

Figure 1. Site location. © Crown copyright and database rights 2021 Ordnance Survey 100019340.

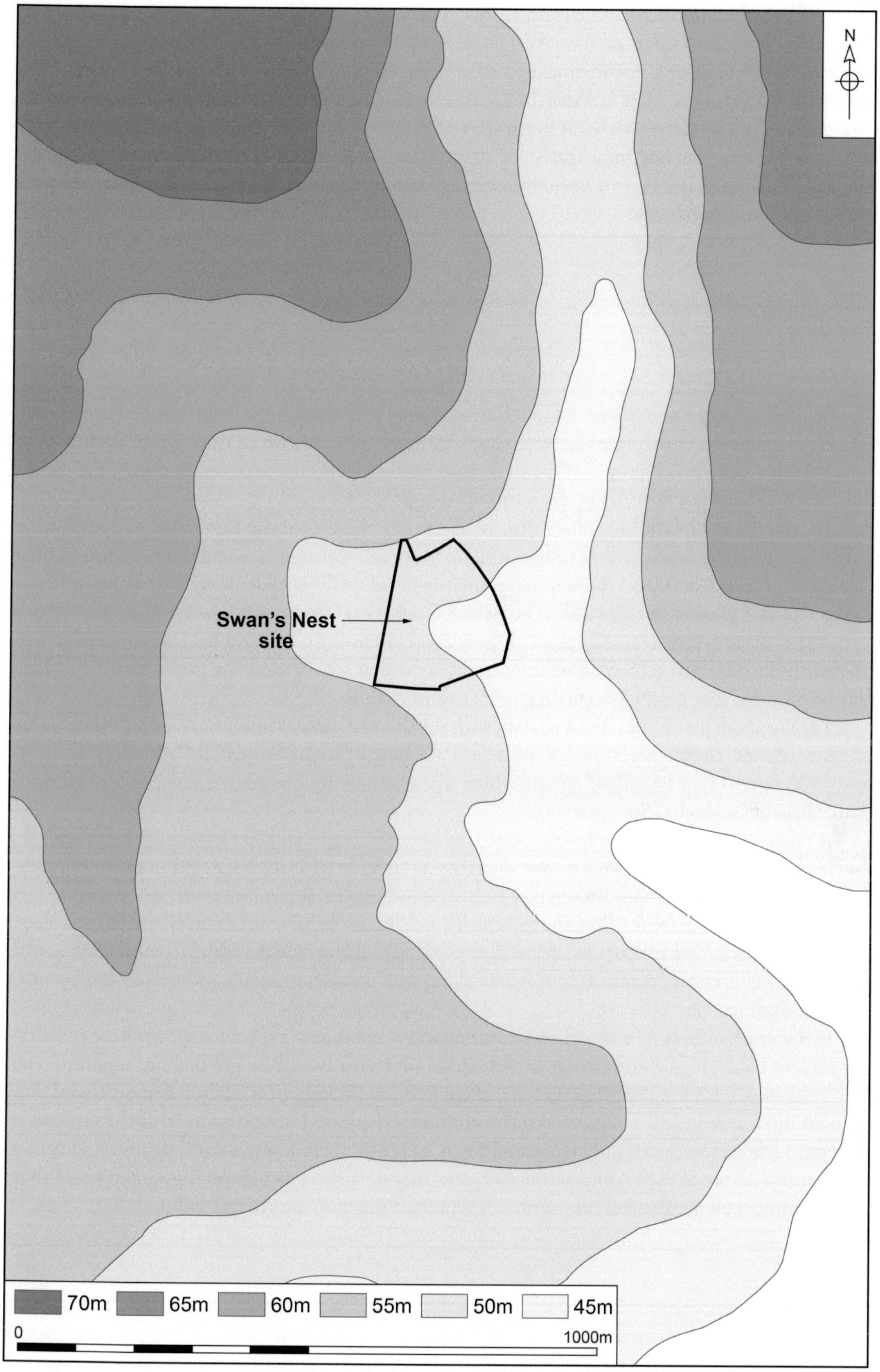

Figure 2. Topographic map of Swan's Nest and environs. © c/o NPS Archaeology.

At Swan's Nest, land rises to the north-east and to the south-west, with Area A sloping gently down from 55.80 m AOD to 49.20 m AOD, and Area B sloping gently down from 54.67 m AOD to 48.90 m AOD (Figure 2). Approximately 1.2 km to the south of the site is a dry valley possibly once forming the head of the River Gadder. Today, this small tributary of the River Wissey originates at a spring near Cockley Cley, *c.*2.46 km south-west of the site. Its course winds through the Breckland landscape for *c.*10.3 km and joins the River Wissey near Oxborough Hythe. The nearest accessible point of the River Wissey is *c.*3.7 km south-west of the site, with the River Nar a minimum distance of *c.*9.6 km to the north-east.

Geology

The bedrock geology underlying Swan's Nest comprises the Lewes Nodular Chalk Formation, a sedimentary deposit that formed between approximately seventy-two to ninety-four million years ago (British Geological Survey 2018). This deposit was exposed on one occasion during the excavation, at a height of 52.3 m AOD in the south of Area B.

The overlying superficial geology present at the site consists of the Lowestoft Till Formation, a sediment glacially deposited up to two million years ago during the Quaternary Period (British Geological Survey 2018). This formation is primarily a chalky till (boulder clay), with occasional outwash sands and gravels, and silts and clays (British Geological Survey 2018). Its variable make-up was reflected at Swan's Nest, where a yellowish-grey clay containing chalk and flint clasts was the most frequently encountered geological sediment, with a distinct area of sand and gravel containing large marine rounded flint nodules occurring in the east of Area B.

The northern part of the Brecks, along with much of its eastern border, is distinguished by the presence of these chalky till (boulder clay) deposits belonging to the Lowestoft Formation. This contrasts with much of the Breckland region, which is predominantly situated upon deposits of sand and chalk (British Geological Survey 2018).

A relatively compact topsoil with a depth of between 0.2 m and 0.53 m was present across the site. It consisted of a dark greyish brown slightly clay-rich silt with abundant sub-angular medium-sized pebbles. A pale orangey brown clay-rich sandy silt upper subsoil measuring between 0.05 m and 0.5 m thick was present below the topsoil. This deposit was however entirely absent in the northern half of Area A – a part of the site more severely affected by modern agricultural activities, with plough scars abundantly recorded in this area along with a shallow topsoil, most likely indicative of horizontal truncation.

In the south of Area A, where the overburden was deepest, lower subsoil deposits were occasionally present consisting of mid-orange brown sandy silts with abundant sub-angular medium-sized flint inclusions, measuring between 0.14 m and 0.4 m thick. Stratigraphic analysis demonstrated these deposits did not represent a single sediment uniformly deposited in one episode, as the same archaeological features were variously truncated-by or sealed-by these lower subsoil deposits. However, the identical nature of their composition indicates that the processes behind the formation of these lower subsoils were similar and they probably represent the remnants of old soils.

Archaeological and Historical Background

Up until very recently, archaeological information on the environs of the site has previously been extremely limited, but the recent southwards expansion of Swaffham through large residential developments has changed this significantly. The discovery of several previously unknown archaeological sites brought to light through pre-development investigations has provided new insights on past human activity in the area. The subject of this publication, the prehistoric site of Swan's Nest, is particularly significant and central to this new understanding.

In addition to Swan's Nest, prehistoric activity has now also been identified just under a kilometer to the south-west and significant Romano-British activity has recently been excavated *c.*200 m to the south of the site. Such discoveries emphasise that the far south-east of Swaffham, an area apparently devoid of notable human activity across the last millennium, appears in contrast to have been a favourable site for occupation throughout prehistory and across the first half of the first millennium AD.

Prior to these recent investigations, our understanding of the prehistory of the area (and the parish of Swaffham more broadly) came almost exclusively from the recovery of chance finds recorded on the Norfolk Historic Environment Register (hereafter NHER). Several Palaeolithic handaxes have been collected from Swaffham, although the exact find locations for most of these artefacts are not known. Only a single Mesolithic find was originally recorded from the parish – a tranchet adze recovered from the north of the town (NHER 13339). Such a pattern stands in contrast to the wider Breckland region, where surface finds of Mesolithic struck flint are frequently recorded (Austin 1997: 9). However, greater quantities of Neolithic material have been recorded from the vicinity of the site. A polished flint axe was retrieved *c.*1 km to the north-west (NHER 2653), with a variety of struck flints also recovered during metal detecting in the surrounds (NHER 25140; NHER 29207; NHER 29209; NHER 41938). A wealth of Bronze Age metalwork was retrieved from the parish during the nineteenth century (including a Middle Bronze Age hoard comprising two spearheads, a chisel, a flanged axehead and a palstave (NHER 2664)), though the find locations of most of these artefacts remains unknown. Apart from a single Late Bronze Age–Iron Age pin found *c.*1.7 km north-west of the site (NHER 37178), no other Iron Age finds had been recorded from the parish prior to the excavation at Swan's Nest.

Romano-British activity was previously only evidenced by a disperse assemblage of artefacts dating from the first to fifth century AD, recovered during metal detecting on agricultural land near the site. This included brooches (NHER 17420; NHER 28567; NHER 36986), coins (NHER 17420; NHER 28566; NHER 29207; NHER 29209; NHER 36986) and steelyard weights (NHER 29207; NHER 51232).

As outlined, findings from recent excavations around the south-east periphery of Swaffham have substantially added to this sparse collection of artefacts. A Late Bronze Age/Early Iron Age pit and two Late Iron Age ditches were recorded during excavations *c.*200 m to the south of Swan's Nest (NHER 41938) (Albion Archaeology 2005; Sleap and Savage 2016: 9–12), and at land west of Brandon Road, *c.*750 m south-west of Swan's Nest, a Middle Bronze Age L-shaped enclosure ditch with associated pits and post-holes was discovered (NHER 64330) (Blackbourn 2021). The Romano-British presence in this area was also characterised through the excavations *c.*200 m to the south of Swan's Nest where a large rectangular building dating from the first to second century AD was discovered, interpreted as a granary (Sleap and Savage 2016). A Middle–Late Romano-British phase was also identified at this site, with corn-dryers, pits and ditches providing evidence of an established rural settlement (NHER 41938) (*ibid.*).

Consequently, evidence now points towards the extensive utilisation and occupation of the south and south-east of Swaffham throughout later prehistory and the Romano-British period. However, the paucity of significant post-Roman activity in this area has also been made apparent through these investigations. It is difficult to ascertain why, after such a long period of human exploitation, activity largely ceased in this area from the Anglo-Saxon period onwards. It is clear, however, that around this time the focus of settlement appears to shift to the core of the modern town. In 1970, an excavation carried out just over *c.*1 km to the north-west of Swan's Nest in the centre of Swaffham investigated part of an Early Anglo-Saxon cemetery, with one cremation and nineteen inhumations identified (NHER 1125) (Hill and Wade-Martin 1976), attesting to the presence of a sixth-century community around the modern settlement core. By the early medieval period, the Domesday Book records Swaffham as a large settlement, with a population of forty-nine households in 1086 (Morris 1984). Activity throughout the medieval period also appears to have been largely restricted to around the modern settlement core. The parish church dedicated to Saint Peter and Saint Paul, parts of which date to the thirteenth century, still occupies the centre of the town (NHER 2698). Remains of a medieval mill mound (NHER 2655) and a possible dwelling (NHER 37599) are also known from the modern settlement core, with medieval pits and post-holes also identified during excavations in this area (NHER 36663; NHER 40197) (Hobbs 2002, 2004).

Evidence of Anglo-Saxon and medieval activity in the vicinity of Swan's Nest is restricted to the occasional recovery of dispersed artefacts through metal detecting, such as coins, brooches and rings (NHER 29205; NHER 29210; NHER 29212; NHER 36987). It is reasonable to consider that such finds result from casual loss and discard during these periods, with only a transient level of activity occurring across what appears to have been a marginal area of the parish. An alternative source for such finds could be from the spreading of night soil during the post-medieval period, and consequently they may originate from outside of the Swan's Nest vicinity entirely.

Post-medieval evidence is also largely restricted to metal artefacts recovered from the agricultural land surrounding the site, with coins, buckles and tokens found (NHER 28566; NHER 29205; NHER 29207; NHER 30822; NHER 51232; NHER 60567). Other than agricultural field boundaries, the only frequent post-medieval sub-surface archaeological features identified during excavations in the area consist of large pits, likely cast for the extraction of marl. A concentration of such features has recently been identified and excavated to the immediate west of Brandon Road (NHER 64330) (Blackbourn 2021). These appear to have been situated close to one of the primary routes into Swaffham, perhaps to allow for quarried material to be easily transported into the town's core. Both Faden's Map of Norfolk published in 1797 and the 1840 Swaffham Tithe Map show the whole southern part of modern Swaffham as not having been settled or developed, consisting almost exclusively of agricultural land organised through a field system developed around Brandon Road and Watton Road.

Previous Archaeological Investigations at Swan's Nest

A requirement for archaeological work at the site was first prompted by Norfolk County Council Community and Environment Services (hereafter NCCCES) issuing a brief for evaluation investigations in 2011, following the submission of a development proposal by Abel Homes Ltd. The development site was situated across four separate arable fields: Area A in the north-east, Area B to the south-east, Area C in the north-west and Area D to the south-west (Figure 3). A desk-based assessment and field walking survey was carried out during September 2011. Field walking and metal detecting over all four areas (a total of *c.*19.6 ha) recorded fifty-eight struck flints, mostly of Early Neolithic date, and thirty-eight fragments of heat-altered (burnt) flint (Quinn and Peachey 2012). A single sherd of Romano-British pottery was also recovered, along with a small collection of medieval and post-medieval finds (*ibid.*). As part of the evaluative investigations, a gradiometer survey was carried out at the site in October 2011 which identified the presence of a large enclosure across

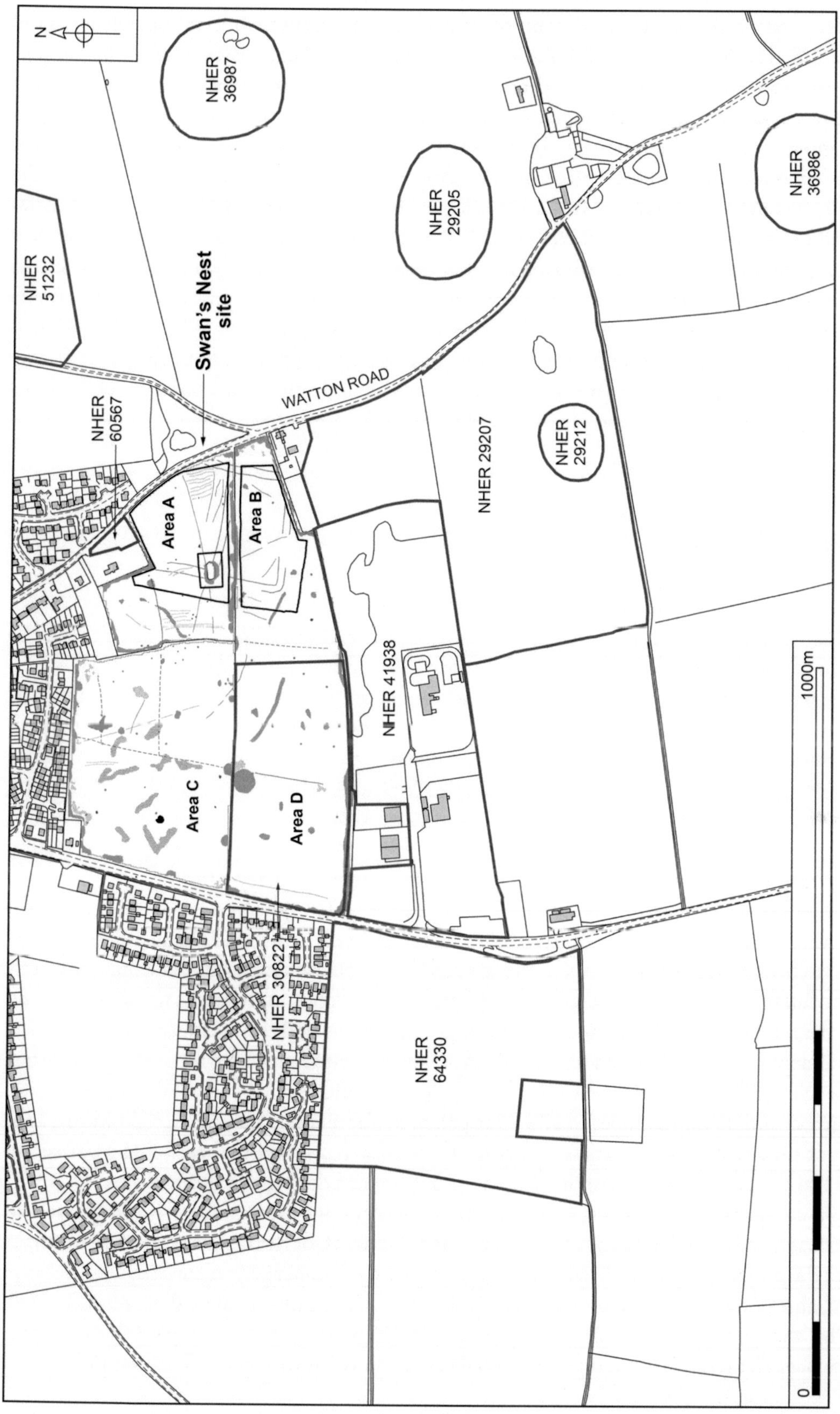

Figure 3. Swan's Nest geophysical survey results (after Marsh 2011) and selected nearby HER events/data. © Crown copyright and database rights 2021 Ordnance Survey 100019340.

Area A and Area B (Figure 3) (Marsh 2011). Several other potential linear features were also identified by this geophysical survey. Following these considerable results, in September 2012 three trial trenches were excavated along the easternmost edge of Area C to confirm that the archaeological remains identified in Area A and Area B did not continue to the west, as was suggested by the geophysical survey (Smith 2012). With only three ditches identified across these trenches (two dating to the post-medieval period and one undated), it was with greater certainty that the archaeological remains of interest could be considered to be confined to Areas A and B. Trial trenching of what was now identified as an enclosure took place in May 2016. Six trenches were targeted to examine the enclosure ditches and test for any internal archaeological features or deposits. A further six trenches were located around the outside of the enclosure. Apart from the double-ditched enclosure, few features of significance were recorded during the trial trenching. The recovery of twenty-two sherds of Late Iron Age–Early Romano-British pottery from the enclosure ditches initially led the authors to suggest the enclosure was of 'Late Iron Age date, with some suggestion that [its] use continued beyond the mid-first century and into the Romano-British period' (Bull and Smith 2016). The results of the trial trenching have not been integrated into this publication – the rationale behind this omittance is presented in the excavation archive report (see White 2020).

Excavation Methodology

In 2017, NCCCES issued a brief for the full excavations of the Swan's Nest enclosure in order to recover as much information as possible on the origins, date, development, phasing, spatial organisation, character, function, status, significance and the nature of social, economic and industrial activities carried out at the site in the past (Hickling 2017). The specified excavation area covered all of the enclosure present within the proposed development site, as well as a small additional buffer zone outside of the ditches. The total excavation area spread across the two fields and measured *c.*4.2 ha. An 8 m wide zone around a large pond present in Area A was not stripped. Following the production of a Written Scheme of Investigation by NPS Archaeology (Adams 2018), the excavation commenced on the 15 October 2018 and was completed by the 15 February 2019.

Topsoil and subsoil deposits were removed in successive *c.*0.1 m spits by a mechanical excavator under archaeological supervision, with excavation halted either when archaeological or geological deposits were encountered. All discrete features were at least 50 per cent excavated, and approximately 10 per cent of all linear features were excavated. Where it was deemed appropriate, in order to obtain additional data, some pits and post-holes were 100 per cent excavated. All excavated archaeological features and deposits were recorded using NPS Archaeology single context recording pro forma. A total of seventy-nine soil samples were taken during the excavation, which included bulk, monolith, pollen, charcoal and diatom samples to allow for paleoenvironmental analysis. Aerial photographs were taken of the site during the excavation using a Zenmuse X4S mounted on a DJI Matrice 210 drone (see Plate 1). An overall plan of the site was also produced during this drone flight.

After the completion of fieldwork, an assessment report detailing the excavation findings was produced, which assessed their significance and outlined requirements for further analyses and a means to disseminate the project results (see White 2019). Subsequently, the specialist recommendations were undertaken, and a detailed stratigraphic analysis of the archive was carried out by the author. The full results of this were presented in an archive report (see White 2020). On the completion of the analysis, all artefacts, paper records and digital materials were deposited with Norfolk Museum Services.

Publication Structure

The archaeological remains identified at Swan's Nest belong to eight different archaeological periods, which are presented in chronological order in Chapter 2. Detailed specialist reports on the different artefact assemblages recovered by the excavation are individually presented in Chapter 3 and discussed by material type and period. In Chapter 4, the faunal and paleoenvironmental evidence is presented by material type and period, with the results of the aerial imagery analysis of the Swan's Nest environs presented in Chapter 5. Chapter 6 interprets the results and discusses the significance of the findings from Swan's Nest and places them within a broader regional context.

The archaeological remains were primarily assigned to periods based on the dating of artefacts and calculated radiocarbon dates. The archaeological periods and their respective dates used in this volume are outlined in Appendix C. In places, this dating evidence has been extrapolated to features where artefacts were not recovered through analysis of spatial and stratigraphic relationships, along with the identification of similarities in the form and fills of features. In places, the certainty of the assigned period of a feature/deposit is not absolute; where this is the case, the caveats and limitations are outlined. This volume fully disseminates the prehistoric remains identified at Swan's Nest, with the geological, undated and post-prehistoric periods summarised. The eight broad phases of activity identified at the site were as follows:

- Period 0: Geological
- Period 1: Mesolithic
- Period 2: Early Neolithic
- Period 3: Middle Bronze Age
- Period 4: Late Bronze Age
- Period 5: Earlier Iron Age
- Period 6: Romano-British
- Period 7: Medieval
- Period 8: Post Medieval
- Period X: Undated

Plate 1. Swan's Nest, Swaffham, looking south-west. © c/o NPS Archaeology.

Where features are considered to have had a shared function or comprise part of the same structure, they have been placed into groups and presented accordingly. The numbering used throughout these groups does not imply a chronological development between separate groups. Throughout this publication, assigned context numbers from the excavation are given in round brackets and sample numbers are distinguished by angle brackets. For figures, cut numbers are given in bold text and deposits are given in standard text, both without brackets.

CHAPTER 2

Excavation Results

Period 0: Geological

The machine-exposed surface geology was dense with periglacial features, mostly concentrated along the margins of the clay till and weathered chalk outcrops. These 'ice-wedge'-like features were linear in plan and mostly aligned north-west-to-south-east. Many of these geological striations were identified on the gradiometer survey previously conducted at the site. To confirm their geological and not anthropogenic origin, several were excavated. The excavated sample had irregular, undulating profiles which frequently undercut the superficial till and outcropping chalk. They were filled with sterile pale yellowish-brown fine sandy silts, which contained rare small sub-angular flint inclusions.

Period 1: Mesolithic

Struck flints typologically dating to the Mesolithic period (10000–4000 BC) were recovered as residual artefacts from later features. The total Mesolithic assemblage from the site consists of a tranchet adze, eight blades and four flakes (see Peachey: this volume). The recovery of a complete undamaged tranchet adze from earlier Iron Age feature Linear 1 (697) is significant and attests to an appreciable hunter-gatherer presence at the site during the Mesolithic.

Period 2: Early Neolithic

Introduction

A discrete area of Early Neolithic (4000–3300 BC) activity was identified at the site in Area B, comprising four post-holes and a distinctive cluster of eleven intercutting pits (Figure 6). A moderate quantity of residually deposited struck flint with typologically Early Neolithic characteristics was also recovered from across the Middle Bronze Age enclosure, particularly from the tertiary fills in Area B (see Peachey: this volume). Two residual sherds of Early Neolithic pottery were also recovered from the Middle Bronze Age enclosure ditches.

The Pit Cluster

A cluster of eleven intercutting pits identified in Area B (Pit Cluster 1), which resembled in plan the shape of a hook or a question mark, were 100 per cent excavated. The cluster was situated on a discrete patch of sandy natural; something which appears to have been intentional when considering the extent of the surrounding dense chalky clay. These pits, though not all identical in character, shared common properties and tended to be of broadly similar shape, form and size. Although there was a moderate degree of variation in their depths (ranging from 0.15 m to 0.57 m), there was greater consistency in their diameters; most being sub-circular in plan and measuring between 0.55 m and

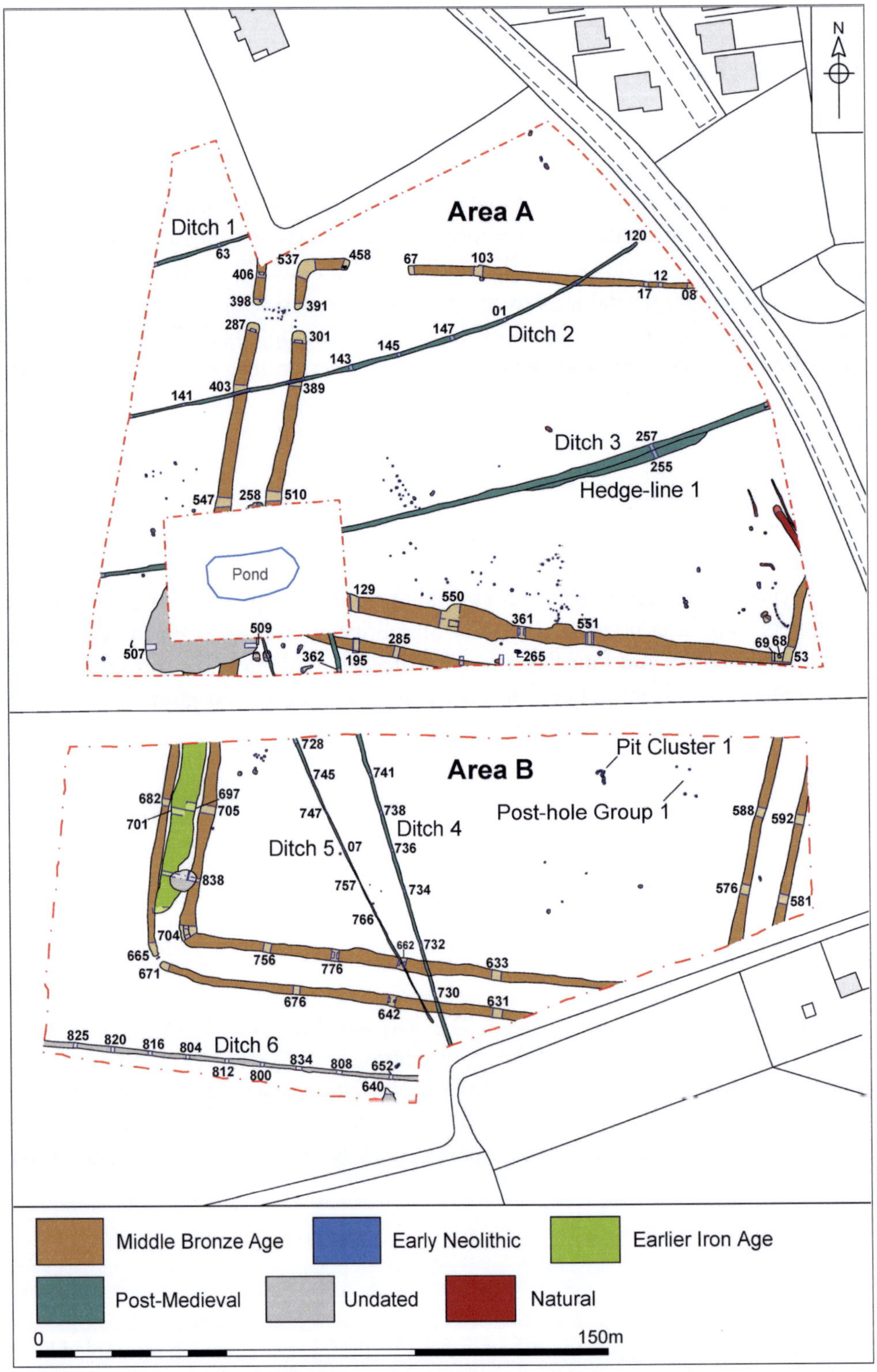

Figure 4. Swan's Nest: Plan of all features, all periods. © Crown copyright and database rights 2021 Ordnance Survey 100019340.

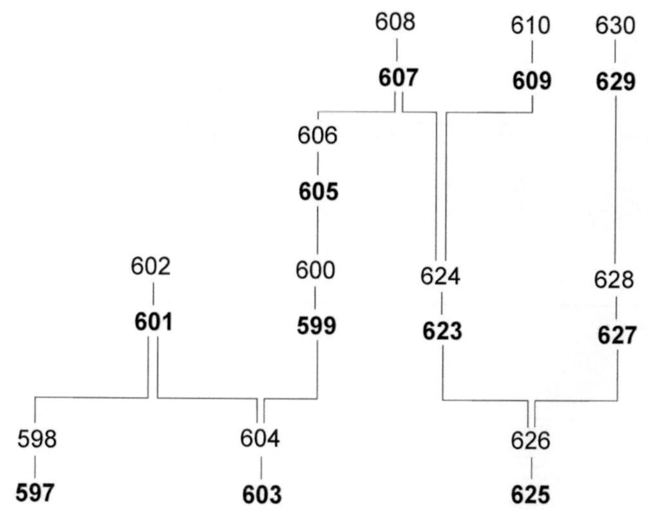

Figure 5. Harris Matrix of Early Neolithic Pit Cluster 1. © c/o NPS Archaeology.

0.74 m. Two of these pits, (623) and (625), were oval in plan. All the pits within this cluster contained single fills, consisting of friable brown silty sands which varied in hue.

A moderate quantity of material culture was recovered from these pits, with a combined assemblage of 123 pottery sherds (1,267 g), forty-seven struck flints (771 g), eleven fragments of animal bone (75 g) and two intrusive pieces of heat-altered flint (22 g). From above the pit cluster, a small assemblage of artefacts was collected during machining and hand-cleaning. This assemblage comprised twelve sherds of Early Neolithic pottery (55 g), two pieces of burnt flint (22 g), two retouched flint end scrapers, a worn flint blade and four struck flint flakes (111 g). Although the exact order in which the pits were dug was not discernible, analysis of their stratigraphic relationships provides some chronological resolution (see Figure 5).

Pit (597) was one of the earliest pits cast in this sequence. It was circular in plan with a slightly concave profile, near vertical sides and a moderately sloping base. It represents one of the most substantial pits in this group, measuring 0.57 m deep and 0.55 m in diameter. This feature was cut by pit (601). Unlike other pits in the cluster, (597) had a shaft-like profile, similar to some post-holes. However, in the absence of a post-pipe or packing materials, there remains uncertainty as to whether this feature does represent a post-hole. It was filled with (598), a mid-greyish brown sandy silt with frequent small sub-angular flint inclusions. It contained an assemblage of twenty-three sherds of Early Neolithic pottery (mostly deriving from bowl-shaped vessels) and six pieces of struck flint including two blades, which both exhibited signs of wear (one piece having possibly been used as a chisel). Hazel nutshell fragments were recovered from a sample of deposit (598), along with charcoal fragments of hazel or alder and oak. The fill of this pit was also sampled for pollen (see Green: this volume).

Pit (603) was also one of the earliest features to have been dug in the cluster. This pit was sub-circular in plan, with a diameter of 0.70 m and depth of 0.15 m. It had been significantly cut away by later pits (599) and (601), which had distorted its profile considerably. Where its profile survived it had moderate sloping sides and a gentle sloping base. It was filled with (604), a mid-orange brown slightly sandy silt with frequent small sub-angular flint inclusions, from which a single sherd of Early Neolithic pottery and a struck flint flake were recovered. Another early feature in the sequence was pit (625), which was the largest pit in the cluster. It was sub-oval in plan with a length of 0.96 m and width of 0.51 m. It had a concave bowl-shaped profile with gently sloping sides and base. It was filled with (626), a homogenous mid-orangey brown silty sand containing occasional small sub-angular flint inclusions. No artefacts were recovered from the fill of this feature.

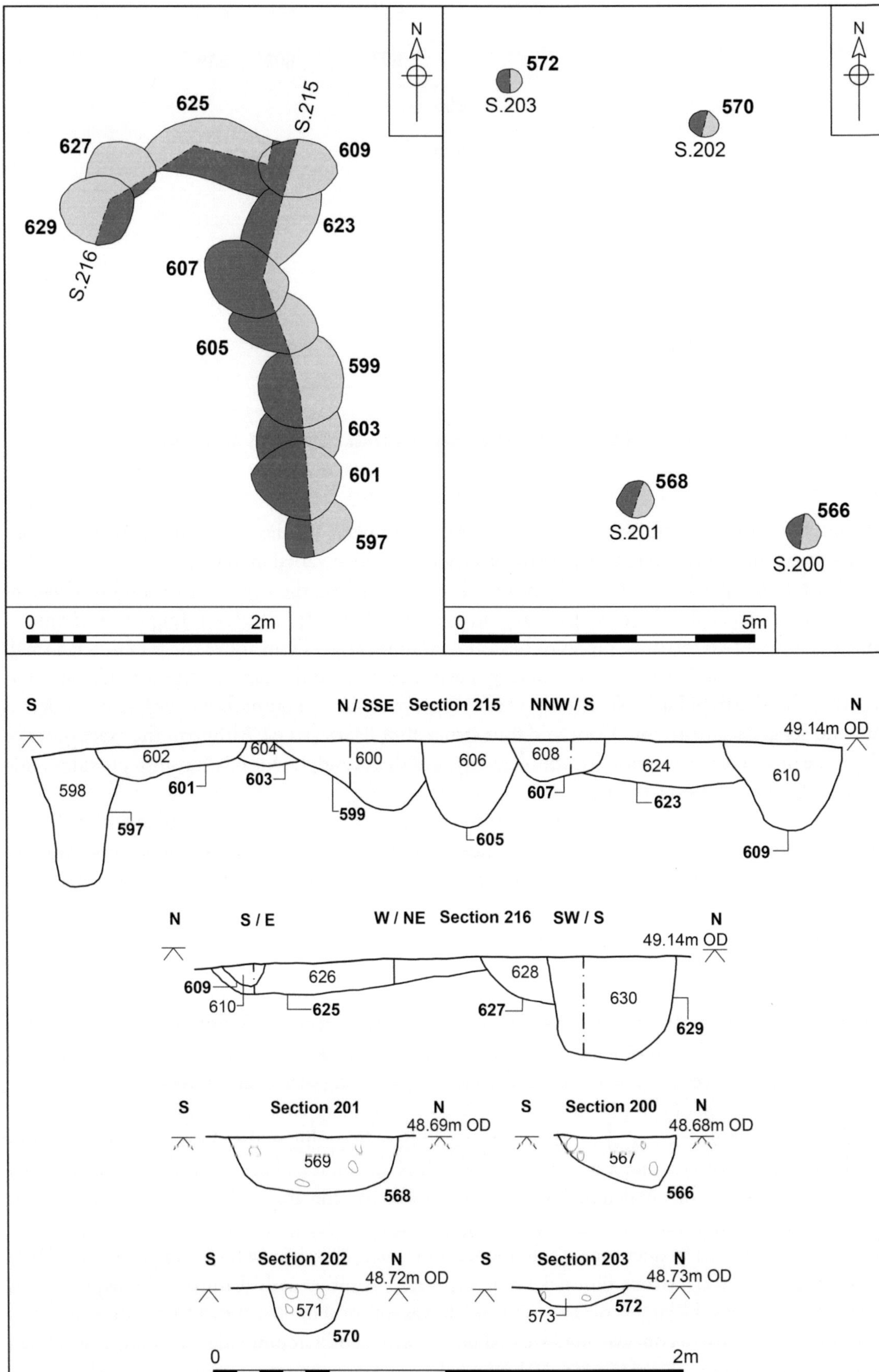

Figure 6. Period 2: Early Neolithic features, plan and sections. © c/o NPS Archaeology.

Following the digging of these three pits, and their backfilling with sand containing fragments of cultural material, further pits were then dug into the same area. Pit (599), which cut pit (603), was sub-circular in plan with a diameter of 0.72 m and a depth of 0.28 m. It had a slightly concave profile with moderately sloping sides and base. This pit was filled with (600), a deposit of dark greyish brown silty sand containing frequent small sub-angular flints. An assemblage of twelve Early Neolithic pottery sherds, derived from cup and bowl vessels, a fragment of cattle tibia and four struck flints including a retouched end-scraper were recovered from this fill. Similarly, pit (601) truncated both pits (597) and (603). It was sub-circular in plan, having a diameter of 0.70 m and a depth of 0.15 m, with moderate sloping sides and a gentle sloping base. Pit (601) was filled with (602), a dark greyish brown slightly sandy silt with frequent small sub-angular flints, from which nine sherds of Early Neolithic pottery, fragmented juvenile cattle teeth and five struck flint flakes were recovered.

Pit (623), which was oval shaped in plan, truncated pit (625). Pit (623) measured 0.80 m long and 0.54 m wide and had a slightly convex profile with gently sloping sides and base. It had a depth of 0.19 m and was filled with friable mid-orangey brown sandy silt (624), which contained occasional small sub-angular flint inclusions. From this fill seven sherds of Early Neolithic pottery (one with Mildenhall-style traits) and a single struck flint flake were collected. In the north-west of the cluster was pit (627), which was circular in plan with a diameter of 0.59 m. It cut pit (625) and was cut by pit (629). Due to its truncation by pit (629) its profile was distorted to the south-west, but to the north-east it had moderate sloping sides and a gentle sloping base, attaining a maximum depth from the machined ground level of 0.19 m. It was filled with deposit (628), a mid-orangey brown silty sand with frequent small sub-angular flint inclusions. Two sherds of Early Neolithic pottery, two flint blades and three flint flakes were recovered from this fill.

Pit (605) was sub-circular in plan with a diameter of 0.57 m. It cut pit (599) and was cut by pit (607). This pit had a 'U'-shaped symmetrical profile measuring 0.38 m deep with steep sloping sides and a moderate sloping base. It was filled with (606), a dark greyish brown sandy silt with frequent small sub-angular flint inclusions, which was sampled for pollen (see Green: this volume). Fill (606) contained the largest individual assemblage of Early Neolithic pottery from the cluster (thirty-one sherds (469 g), some of which had notably thick rims). Three retouched end scrapers and two flint flakes were recovered close to the base of this feature. A quartzite cobble with no distinct signs of anthropogenic modification was also collected from deposit (606). Hazel nutshell fragments were recovered from sample <48> of this deposit.

By the time some of the last pits in the cluster, (607) and (609), were dug, at least six earlier pits within a *c.*2 m radius had already been dug and backfilled. Pit (607) was dug into the centre of the cluster where it cut pits (605) and (623). It had a bowl-shaped profile measuring 0.17 m deep, with steep sloping sides and a base which gently sloped to the south-east. It was sub-circular in plan with a diameter of 0.74 m and was filled with deposit (608), a mid-orangey brown silty sand containing frequent small sub-angular flint inclusions. It contained a finds assemblage consisting of a retouched flint side-end scraper, two struck flint flakes and eleven sherds of Early Neolithic pottery. Pit (609) was a sub-circular in plan feature with a diameter of 0.62 m, and cut both pits (623) and (625). It had a depth of 0.35 m, with an asymmetrical profile consisting of a concave south-east edge and convex north-east edge. Its sides sloped steeply to a flat base. This pit was filled with deposit (610), a sandy silt of a dark brown hue which contained moderate quantities of sub-angular small flint inclusions. From this pit, eight sherds of Early Neolithic pottery derived from bowl-shaped vessels and five struck flint flakes were recovered. A possible cotyledon fragment of an indeterminate large pulse (*Fabaceae*) was recovered from sample <49> of this deposit, along with hazel nutshell fragments and pieces of oak charcoal. The cotyledon fragment of *Fabaceae* returned a radiocarbon date of 3645–3521 cal. BC (Beta – 552730, 95.4 per cent).

Pit (629) was also one of the last pits dug into the cluster, truncating pit (627). This feature was circular in plan with a diameter of 0.63 m and symmetrical 'U'-shaped profile with steep sides and a gently sloping base. It had a depth of 0.42 m and was filled with (630), a mid-greyish brown silty sand with frequent small sub-angular flint inclusions. It contained eleven sherds of pottery and six struck flint flakes. Hazel nutshell fragments were recovered from sample <50> of this deposit.

Post-Hole Group 1

Approximately 20 m to the east of Pit Cluster 1 were four similar post-holes set in a rhomboidal arrangement, considered to be of Early Neolithic date. Post-hole Group 1 consisted of individual post-holes (566), (568), (570) and (572).

Post-hole (566) was circular in plan with a diameter of 0.50 m and a depth of 0.20 m. It had a 'U'-shaped symmetrical concave profile with steep sloping sides and a gentle sloping base. This feature was filled with deposit (567), an orange brown silty sand with occasional small sub-angular flint inclusions. No finds were recovered from the fill of this feature. Approximately 2.20 m to the east of post-hole (566) was post-hole (568). It was circular in plan with a diameter of 0.52 m and depth of 0.22 m. It had a 'U'-shaped symmetrical profile with steep sloping sides and a gentle sloping base, and was filled by (569) a yellowish-brown silty sand with occasional small sub-angular flints. No finds were recovered from this feature.

Post-hole (570) was situated *c.*5.60 m north of post-hole (568). This feature was circular in plan with a 0.50 m diameter and depth of 0.18 m. It had a symmetrical profile with moderate sloping sides and a gentle sloping base. It contained deposit (571), a mid-brown sandy silt with occasional small sub-angular flint inclusions and rare charcoal flecks. It is possible that the charcoal flecks and fired clay indicate that hearth waste was a component of the deposit that filled this feature. Two tertiary flint flakes were recovered from deposit (571) along with ten fragments of fired clay. Post-hole (572) appeared to be paired with post-hole (570), which was situated *c.*2.80 m to the east. This post-hole had a 'U'-shaped symmetrical profile with steep sloping sides and a gentle sloping base. It was filled by (573), a deposit identical in composition to (571). From this feature an assemblage of five sherds of Early Neolithic pottery, four struck flint flakes and two struck flint chips were recovered.

Although dateable artefacts were only retrieved from the fill of post-hole (572), their general similarities and arrangement strongly suggest they were contemporary. It is likely that they represent the remains of a fence-line; however they may equally have been structural in nature.

Period 3: Middle Bronze Age

Introduction

The most significant period of activity dated to the Middle Bronze Age (1600–1200 BC), with the creation of a double-ditched enclosure at the site (Figure 7). In Area A, distinct deposits indicative of 'burnt mound' activity being carried out inside the enclosure during this period were present. Several post-hole groups and a number of discrete pits were also identified inside the enclosure.

The Double-Ditched Enclosure

The excavation exposed an enclosure of rectangular shape in plan, which was aligned NNE-to-SSW. It was set out on land which sloped gently north-to-south from *c.*55.80 m AOD to *c.*48.90 m AOD, rising again in the same manner further to the south, to *c.*54.67 m AOD. Its centre was present across the lowest point, with two approximately equal-sized compartmented areas juxtaposed to the north-east and south-west respectively.

The archaeological features defining the enclosure were ditches and a total combined length of approximately 840 m of (non-continuous) ditch was exposed, of which just over 10 per cent was

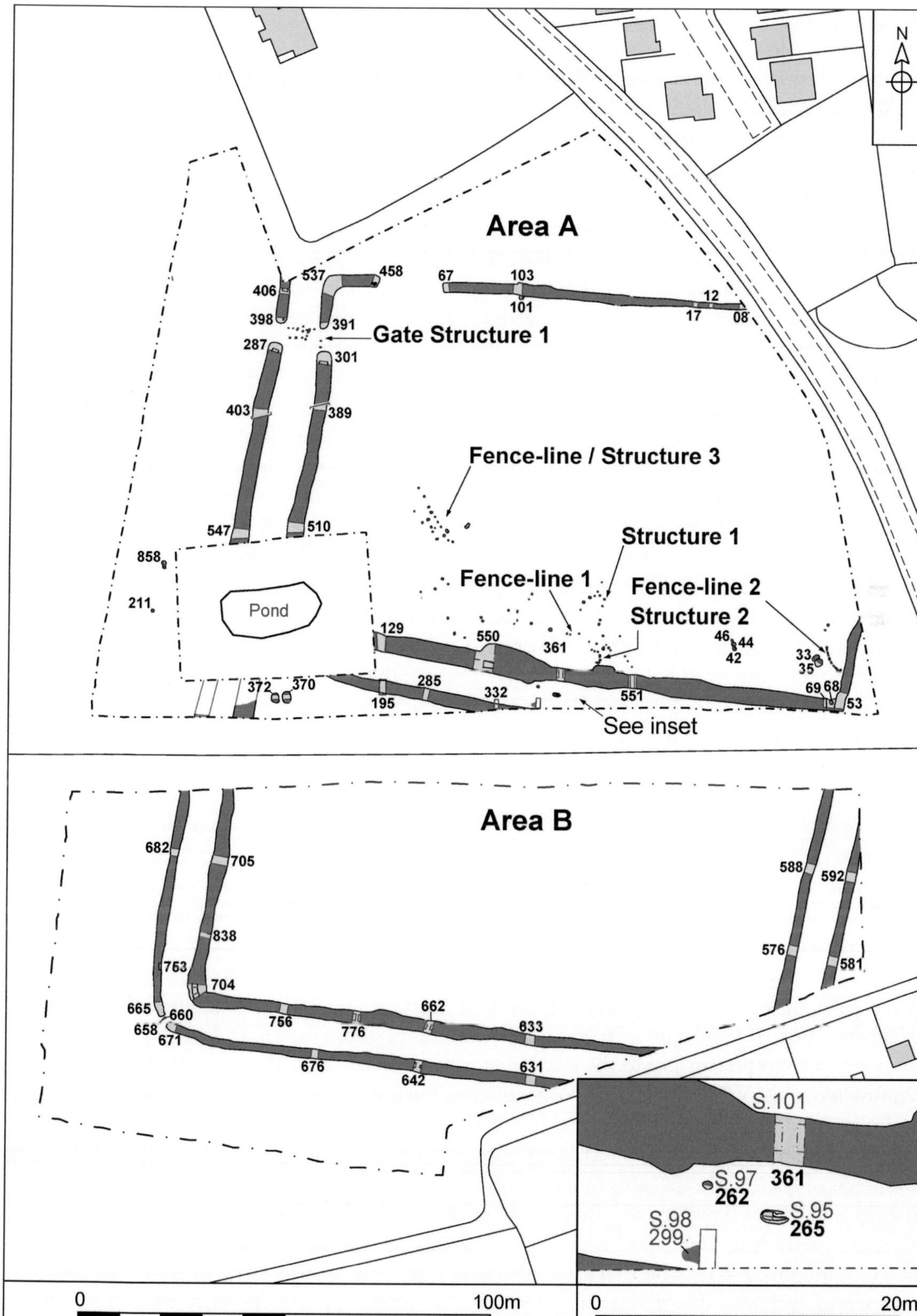

Figure 7. Middle Bronze Age features. Inset: detail of 'burnt mound' activity area. © Crown copyright and database rights 2021 Ordnance Survey 100019340.

hand-excavated. These excavated slots were typically 2.00 m in length and positioned broadly equidistant, with allowances in their spacing made to investigate specific areas of interest (e.g. termini, corners and other unique elements). With the exception of the north-east extent, the enclosure was entirely of a double-ditched form, with the space in-between ditches varying from 6.75 m to 10.95 m in width. As previously mentioned, the enclosure was split in to two approximately equal parts; the enclosed area to the north is projected as having measured approximately 11,500 m^2 (*c.*8,680 m^2 of which was investigated in Area A) and the enclosed area to the south is projected as having measured approximately 9,750 m^2 (*c.*7,860 m^2 of which was investigated in Area B, and *c.*115 m^2 was investigated in Area A). A *c.*1,500 m^2 area in the south-west of Area A was not excavated due to a buffer zone being left in place around an extant pond. The enclosure appears to have extended to the north of the site, with one of the ditches continuing underneath the south-east facing baulk in Area A.

The average depth of the ditches forming the enclosure (from the machine stripped natural horizon) at the break of slope (top) was 1.17 m, with a range of 0.30 m–1.75 m. The average width (at the top of the break of slope) was 2.61 m, with a range of 0.88 m–7.07 m. It is clear that the recorded width and depth was affected by horizontal truncation at the site, most likely caused by modern agricultural activities and consequently these values cannot be taken as accurately detailing the original size of the ditches when first cast in prehistory; they most likely would have been larger. Horizontal truncation was particularly apparent in the north-east of the enclosure in Area A, where the smallest ditch depth and width was recorded. This truncation has also been taken into account when analysing the fill sequences presented below, and in distinguishing between primary, secondary and tertiary fills. If the prehistoric soil overburden at the site was similar to that of today, an approximate average minimum depth for the enclosure ditches is estimated to have been in the region of 2.20 m. The profile of the enclosure ditches varied across the site, but overall they had well-defined breaks of slope (top and base), moderate sloping sides and flat or gentle sloping bases (see Figure 8 for example sections).

A total assemblage of forty-three sherds (698 g) of Middle Bronze Age pottery was recovered from the deposits infilling the enclosure ditches; the majority of this material being recovered from primary and secondary fills. Radiocarbon dating carried out on charcoal and animal bone recovered from the ditches further supports a Middle Bronze Age date for the casting and primary use of the enclosure (see Table 1 and Appendix D).

Three clear entrances/exits were identified in the parts of the enclosure revealed by the excavation; one appearing to be a significant route into the site in the north-west corner of Area A and the other two representing internal entrances/exits to access different parts of the enclosure. The primary entrance/exit was marked by a termination in four of the ditches in the north-west of Area A (see Plate 2), and an associated arrangement of eighteen post-holes which are discussed below (Gate Structure 1). These appear to demarcate an access passage *c.*2.00 m wide. A probable internal entrance/exit was present just to the north-east of Gate Structure 1. This was marked by a *c.*15.5 m termination in the WNW-to-ESE enclosure ditches. No post-holes indicating the presence of a gate or similar were identified.

A second narrow probable internal entrance/exit with a width of *c.*2.00 m was identified in the south east of Area A. This possibly functioned as a routeway between separate parts of the enclosure in Area A and Area B respectively. In this part of the site, small hollow (68) was present which appears to represent the remnants of a cutting made for ground consolidation, across what was possibly a high-traffic area. It is possible that across this area, timbers or an alternative organic consolidation material was laid to solidify the ground. In this area, post-hole (71) was identified which had a diameter of 0.85 m, a depth of 0.59 m and was filled with deposit (72), a mid-yellowish-brown silty clay with occasional medium-sized rounded flint inclusions. A deposit of dark brown clay-rich silt (73) was present above deposit (72), and may represent the remnants of a post-pipe. The function of this feature is not clear, but it may represent a setting-out post put in for the construction of the enclosure, or the remains of a feature which at some point was used to block access.

Table 1. Radiocarbon age determinations from the Swan's Nest enclosure

Context	Sample code	Uncalibrated result (BP)	Calibrated result (BP)	Calibrated result (BC)
(106)	Beta – 552731	3,160 ± 30 BP	3450–3272 cal. BP	1501–1323 cal. BC
(376)	Beta – 552729	3,140 ± 30 BP	3446–3254 cal. BP	1497–1305 cal. BC
(388)	Beta – 552728	3,120 ± 30 BP	3399–3240 cal. BP	1450–1291 cal. BC
(461)	Beta – 552732	3,010 ± 30 BP	3335–3077 cal. BP	1386–1128 cal. BC

Plate 2. Excavation of enclosure ditch termini in Area A, looking south-east. © c/o NPS Archaeology.

It is unlikely that these represent the only entrances/exits that existed across the enclosure, with others highly likely to have been present in the parts of the enclosure not subject to excavation.

A small opening, 2.10 m wide, was present in the south-west of the outer enclosure ditch in Area B. In this space, two post-holes were identified. Post-hole (658) was circular in plan and had a diameter of 0.23 m, a depth of 0.11 m and was filled with dark greyish brown clay-rich silt (659), with rare small sub-angular flint inclusions. Approximately 0.3 m to the north-east of this feature was post-hole (660), which was circular in plan with a diameter of 0.27 m, a depth of 0.13 m and was filled with dark greyish brown clay-rich silt (661). With the presence of a bank in the area between the two enclosure ditches (discussed below) and no corresponding break present on the inner enclosure ditch, it seems unlikely this represents an entrance/exit into the enclosure. However, this is not conclusive, with the potential for a bridge of sorts to have possibly existed over the inner enclosure ditch. As this was the only corner of the outer enclosure circuit exposed during the excavation, it is entirely possible that this termination was part of the enclosures regular form and does not represent a zone with a particular function. The post-holes identified in this area could represent initial setting-out posts put in during the construction of the enclosure.

The Fill Sequence

The fill sequence across the enclosure consisted of a variety of different deposits that can be grouped into several distinct categories:

- *Type I* – deposits formed from the accumulation of washed-in and water-lain fine silts from the immediate environment around the ditches
- *Type II* – redeposited superficial geological deposits originating from up-cast material created from earlier intrusive digging activities (i.e. mass wasting of the bank)
- *Type III* – accumulated or dumped deposits with a notable organic component (charcoal) and a high silt content, derived from anthropogenic activities
- *Type IV* – tertiary leached soil-like, silt-rich deposits originating from the gradual movement and weathering-in of surrounding soils
- *Type V* – remobilised wind-blown sands
- *Type VI* – deposits consisting almost entirely of fractured burnt flint

To broadly summarise the fill sequence across the enclosure ditches as a whole, the primary fills were dominated by *Type I* and *Type II* deposits, considered to result from gradual weathering, washing and slumping of material from the immediate environment around the ditches. These fills contained Middle Bronze Age pottery and animal bone (with greater quantities recovered from Area A), suggesting a degree of sustained activity within the enclosure at this point. However, no deposits were considered indicative of intense occupation among the primary fills. Analysis of the position of the *Type I* and *Type II* deposits within the enclosure ditches suggest that banks were present across the whole of the enclosure, in the space between the two ditches.

The same slumping and weathering infilling processes continued to occur throughout the life of the enclosure, with *Type I* and *Type II* deposits also recorded among the secondary fills of the ditches. In one part of the enclosure, a discrete area of intense 'burnt mound' activity was evidenced by the presence of several concentrated burnt flint deposits (*Type VI*).

Type III silts with a high charcoal component were also identified among the secondary fills. The precise origin of the *Type III* deposits is not clear, but they were not artefact-rich, possibly suggesting they did not form through accumulations of domestic refuse associated with general occupation. It is more likely that their formation was related to the 'burnt mound' activity identified, with associated combustion waste dumped across the site. These *Type III* deposits may have also been partially composed of excrement from animals stocked within the enclosure.

The 'burnt mound' deposits dated to the Middle Bronze Age, along with most of the *Type III* deposits. Some of these *Type III* deposits however also appear to have been of Late Bronze Age and earlier Iron Age date, and thus may be associated with the later reuse of the enclosure (these are still described in this section for completeness).

The tertiary deposits were largely the same across the entire enclosure and were comprised of *Type IV* silts, most likely originating from the gradual movement and weathering-in of surrounding soils, which then subsequently leeched. In Area B, these deposits contained a significant quantity of residual Early Neolithic struck flint.

Little evidence was provided for the deliberate backfilling/levelling of the enclosure ditches in any period. Artefact and radiocarbon dating evidence suggests that different parts of the enclosure infilled at different rates, taking, it would appear, approximately 1,500 years for the negative earthworks to become completely obscure. Conversely, it is not possible to precisely estimate how long the earthwork banks were visible for at the site, but they were certainly absent by the nineteenth century. No recutting of the enclosure ditches appears to have taken place and evidence of clearing/cleaning or maintenance could not be identified.

Area A: North-East

Six slots were excavated into the north-east arm of the enclosure in Area A (contexts (08=12=17=67=103=123)), with various deposits identified infilling the ditch in this location. In all of these excavated slots, mid-yellowish-brown fine sandy silts containing rare small sub-angular flint inclusions were present infilling the base of the enclosure ditch (deposits (07; 11; 13; 15; 81; 82; 86; 89; 90; 91; 104; 105; 106=108; 124; 125)). These primary deposits are considered to represent washed-in silts from the immediate environment around the ditch (*Type I*). The small assemblage of finds recovered from these fills includes six sherds of Middle Bronze Age pottery (mostly deriving from urns), animal bone, burnt flint, two pieces of fired clay and a piece of burnt carstone, indicating some degree of activity in this area of the enclosure. A cortical fragment of large mammal bone from deposit (106) at the base of slot (103) returned a date of 1501–1323 cal. BC (Beta – 552731, 95.4 per cent).

Two distinct types of deposits constituted the secondary fills in this part of the enclosure. One type comprised organic-rich deposits of friable dark greyish brown sandy silts with frequent charcoal flecks and small sub-angular flint inclusions (contexts (14; 92; 93; 98; 99; 114; 115)). A distinct deposit of friable, dark red fired clay, (118), which contained oak charcoal was also identified. Despite these deposits not being particularly finds-rich (only containing a small assemblage of burnt flint and animal bone), they are considered most likely to originate from either single dumping episodes or the gradual accumulation of organic-rich refuse/material (*Type III*). As these deposits were all tipped in an uphill direction from the south, they are significant in confirming that anthropogenic activities were being carried out within this part of the enclosure.

The other type of secondary deposits recorded in this part of the site comprised compact mid-brownish yellow silty clays with abundant chalk inclusions (deposits (09; 74; 94; 95; 97; 107; 109; 111; 112; 113)). It is suspected that these fills represent the first instances of bank erosion, slumping and collapse (*Type II*) – the bank up until this point appearing to have been relatively stable in this part of the site. From the position and gradient of these deposits within the ditch, it is apparent a bank was present to the north of this length of the enclosure ditch. It is considered unlikely these deposits represent deliberate backfilling episodes, as none of them sealed the ditch entirely or were of a thickness inferring a concerted effort to level the bank or infill the ditch. A small collection of finds recovered from these deposits include animal bone, burnt flint and a sherd of earlier Iron Age pottery, which may indicate these deposits date from a few centuries after the initial Middle Bronze Age activity at the enclosure. Equally, this single sherd may have intrusively entered earlier deposits through bioturbation.

The final tertiary fills in this part of the enclosure were composed of friable dark greyish brown sandy silts with moderate small sub-angular flint inclusions (deposits (06; 10; 100; 117; 119; 126)). These deposits are suspected to have accumulated in the ditch through the gradual movement and washing-in of surrounding soils (*Type IV*). Although it is possible that these deposits were intentionally backfilled into the ditch, this is considered unlikely, as the more preferable material to use in this instance would arguably have been the up-cast deposits which formed the bank. Four sherds of Middle Bronze Age pottery and one sherd of earlier Iron Age pottery were recovered from these fills, along with burnt flint, animal bone and two pieces of fired clay.

Area A: North-West

A total of nine slots were excavated into the enclosure ditches in the north-west of Area A around where the two entranceways were identified (contexts (391=458=537; 301=389; 287=403; 398=405)) (Figure 8). The fill sequence of the enclosure ditches in this area was more varied than in the north-east and artefacts were more abundant. The primary fills of the enclosure ditches in this area comprised both mid-brownish yellow silty clays with abundant chalk inclusions (deposits (289; 399; 406; 408; 412; 561; 695; 696)), most likely representing the weathering and slumping of up-cast natural from the bank soon after the ditches were dug (*Type II*), and mid-yellowish-brown fine sandy silts with rare small sub-angular flint inclusions (deposits (288; 329; 355; 356; 359; 392; 393; 394; 395; 407; 413; 420; 421; 560)), probably representing washed-in silts from the immediate environment around the ditch (*Type I*). A moderate assemblage of finds was recovered from these deposits which consisted of nine sherds of Middle Bronze Age pottery, fired clay, a flint hammerstone and animal bone. A single sherd of earlier Iron Age pottery (suspected to be intrusive) was recovered from deposit (406) and two sherds of residual Early Neolithic pottery were collected from deposit (696).

Secondary infilling deposits also varied significantly across this part of the enclosure. Further sterile deposits of chalk-rich clay bank slump (contexts (358; 360; 414; 415)) (*Type II*) and washed-in water-lain silts were recorded (contexts (290; 328; 409; 422)) (*Type I*). Two sherds of Middle Bronze Age pottery, animal bone and burnt flint were recovered from these deposits, along with a fragment of a limestone anvil from deposit (415) of ditch terminus (391). Several organic-rich deposits were also identified which consisted of greyish brown sandy silts with frequent charcoal flecks identified as deriving from *Prunus* species, oak and alder/hazel (contexts (396; 410; 459; 460)) (*Type III*). Such deposits are recorded regularly amongst the secondary fills of the enclosure ditches in Area A, and as stated previously, likely originate from either single dumping episodes or the gradual accumulation of organic-rich activity-related refuse/material. A finds assemblage consisting of animal bone, burnt flint and a significant collection of pottery was recovered from these deposits. The pottery collected included seven sherds of Middle Bronze Age date from (460) and forty-four sherds of earlier Iron Age date from (396) and (410). Above deposit (459), and sealed by (462), an articulated skeleton of a juvenile cattle SK461 was found, which returned a radiocarbon date of 1386–1128 cal. BC (Beta – 552732, 95.4 per cent).

Through studying the location and form of the bank-slump deposits within the primary and secondary fills described above (specifically deposits (289; 360; 414; 415)), it is apparent that a bank occupied the space between the two ditches of the enclosure in this area. A bank also appears to have been located to the north of ditch terminus (458). This is in keeping with the evidence from the north-east area of the enclosure ditch, where a bank appears to have run along its northern side. Such banks would have been created during the initial excavation of the enclosure ditches, through piling up of the chalk-rich clay up-cast.

The final tertiary fills in this part of the site are similar to those seen across the whole of the enclosure. These deposits comprised friable dark greyish brown sandy silts with moderate small sub-angular flint inclusions (contexts (291; 330; 331; 400; 411; 416; 423; 462; 538)) (*Type IV*). Finds from these deposits included five sherds of earlier Iron Age pottery, one struck flint flake, burnt flint, animal bone and fired clay. Later intrusive finds from these deposits included a sherd of late medieval transitional ware and a post-medieval button.

Area A: South-West

Due to the presence of the extant pond and the unstripped buffer zone surrounding it, only two slots, (510) and (547), were excavated across the enclosure ditches in the south-west of Area A. Due to the height of groundwater during the excavation of these slots, the primary deposits were not reached by hand excavation, with the depth of the features ascertained here through hand-auger. In the outer enclosure ditch (510), only two deposits were recorded; (549), a mid-brownish yellow silty clay with frequent chalk inclusions and (548), a mid-yellowish-brown sandy silt with occasional chalk and small sub-angular flints. The presence of the chalk-rich clay situated on the eastern side of the ditch, representing a redeposited natural, most likely comprises an episode of bank-slumping

(*Type II*). This further supports the presence of a bank in the space between the two ditches of the double-ditched enclosure in this part of the site. The other fill identified, deposit (548), is consistent with the tertiary fills recorded across the enclosure, most likely formed through the gradual accumulation of the surrounding soils into the ditch (*Type IV*). From this deposit, four sherds of earlier Iron Age pottery were recovered along with animal bone and a struck flint flake.

In slot (510), two distinct deposits were recorded. Primary/secondary *Type I* mid-brown sandy silt deposits with occasional chalk inclusions and rare small sub-angular flints (contexts (541; 542; 543; 544)), which appeared to have entered the ditch from the east, represented the lower fills. A small assemblage of animal bone was recovered from deposit (544). These lower fills were overlain by tertiary *Type III* deposits (545; 546), comprised of dark brownish grey sandy silts with chalk and frequent small sub-angular flint inclusions. The charcoal in these deposits derived from hazel/alder, birch and *Prunus* species. Five sherds of Late Bronze Age pottery, animal bone and four pieces of struck flint including a blade were recovered from these fills.

Area A: South and South-East

In the south and south-east of Area A, five slots were excavated into the northernmost ditches (contexts (53; 68; 69=129=361=550=551)), and three slots were excavated into the southernmost ditch (contexts (195=285=332)). The unique fill sequence and findings from (361) are omitted from this section and instead presented in the next.

Across the northernmost ditch in this part of the enclosure, the primary fills entirely consisted of *Type II* deposits composed of compact mid-brownish yellow silty clays with abundant chalk inclusions, most likely representative of initial bank erosion, slumping and collapse which would have occurred shortly after the casting of the enclosure (deposits (54; 55; 56; 75; 76; 130; 552)). From these fills, animal bone and five sherds of Middle Bronze Age pottery deriving from urns were recovered. In slot (550), the basal fills were not reached due to the encroachment of groundwater. A clear pattern was also observed with the secondary fills along this stretch of the enclosure. *Type I* deposits consisting of light to mid-brown clay-rich silts with occasional chalk inclusions and rare small sub-angular flints were most abundant overlying the *Type II* deposits (contexts (57; 131; 132; 135; 294; 295=419; 540; 553; 554; 555; 556; 564)). These deposits appear to have washed into the ditch downslope from the north. From these fills, only animal bone and burnt flint was recovered. Overlying these sterile silts, a group of more organic-rich silts were identified which had a dark brown hue with frequent chalk inclusions (deposits (58; 78; 136; 557)) (*Type III*). An assemblage of four sherds of Middle Bronze Age pottery deriving from urns, along with animal bone and burnt flint was recovered from these deposits.

The tertiary fills in this part of the enclosure were represented by two different deposits. In two of the sections excavated, a further episode of bank erosion, slumping and collapse was identified, made up of pale to mid-brownish yellow silty clays with abundant chalk inclusions (*Type II*) (deposits (59; 60; 61; 137)). One particularly dense deposit, compact and without leeched silts, appears to represent an episode of backfilling of the bank; the only such instance identified across the enclosure with exception of the deposits in slot (361) (see below). This deposit however does not represent an intention to entirely fill the ditch and may instead result from the clearance of a small area of up-cast material which formed the bank. The only finds recovered from these deposits consisted of sparse burnt flint. Across the full extent of this ditch (with the exception of (361)), a final series of deposits was present consisting of friable dark brown sandy silts with frequent sub-angular small flint inclusions (contexts (62; 79; 138; 295; 558; 559)), similar to the final infilling deposits recorded across the entire enclosure (*Type IV*). From these tertiary fills, an assemblage of two sherds of earlier Iron Age pottery, a flint end scraper, a flint blade, animal bone and burnt flint was collected.

Hollow (68), which connected the north-west-to-south-east length of the enclosure ditch with the north-east-to-south-west length of the enclosure ditch in the east of this area, was filled with (69), a mid-orangey brown clay-rich silt with rare small sub-angular flint inclusions. The mechanisms behind the formation of this deposit are not clear – suggestions as to the hollows primary function have been discussed above. A single sherd of Middle Bronze Age pottery deriving from an urn was recovered from this deposit.

In the southernmost ditch along this part of the enclosure, primary fills consisted of *Type II* deposits (194; 284); mid-brownish yellow silty clays with abundant chalk inclusions. As with the northernmost ditch, these likely represent the slumping of bank deposits shortly after the enclosure ditch was cast. Two different secondary deposits were identified in this part of the enclosure. *Type III* deposit (193), a dark greyish brown silt with occasional small sub-angular flint inclusions was present in slot (195), and three yellowish-brown deposits of clay-rich silt, which can be characterised as *Type I* deposits, were recorded in (285) and (332) (deposits (283; 286; 333)). The tertiary deposits present along this length of the enclosure were composed of dark brown sandy silts with rare small sub-angular flint inclusions, likely representative of deposits which accumulated in the ditch through the gradual movement and washing-in of surrounding soils (*Type IV*) (contexts (192; 282; 334)). A single fragment of animal bone from deposit (192) was the only find recovered from the southernmost ditch sectioning the enclosure in two.

From analysis of the *Type II* deposits of slumped redeposited natural in this area of the enclosure ditch (specifically deposits (54; 56; 59; 60; 61; 76; 130; 137; 194; 284)), it is evident that a bank was present between the two ditches that make up the double-ditched enclosure in this part of the site. The direction of mass wasting was primarily from the south-west, but also from the south-east, in the south-east corner of the enclosure in Area A.

Area B: East

Four slots were excavated into the east of the enclosure in Area B, with two excavated into the inner ditch (576=588) and two excavated into the outer ditch (581=592). In this part of the site, the superficial geology was distinctly of sand and gravel containing large marine rounded flint nodules, and was reflected in the composition of the deposits infilling the enclosure ditches in this part of the site. Most of the primary fills in the ditches in this area were represented by *Type I* deposits consisting of mid-greyish brown silty sands with frequent small sub-rounded flint inclusions (contexts (584; 585; 589; 590; 593; 594)). However, in slot (576), a distinct deposit of mid-brownish yellow silty sand (577) was present with a significant quantity of small-medium sub-rounded flint inclusions which represented the primary fill. This deposit is considered to represent a slump of the material which formed a possible bank (*Type II*). The small depth of this deposit in the base of the ditch (0.10 m) does not however inform an understanding of the location of the bank in this part of the enclosure. From these primary deposits, two struck flint flakes and a pottery sherd only dateable as 'prehistoric' was recovered.

Secondary fills were dominated by similar yellow silty sand deposits with significant quantities of small sub-rounded flint inclusions (contexts (583; 595; 611)), most likely deposits of slumped bank material (*Type II*). As with fill (577), their thickness and position does not conclusively identify the location and extent of the bank in this part of the enclosure. A single thumbnail flint scraper was recovered from deposit (595). In slot (576) excavated on the inner ditch of the enclosure, two distinct deposits were present. Above the primary bank slump, deposit (578), a greyish brown sandy silt with occasional small sub-angular flint inclusions and occasional charcoal flecks was present, from which an assemblage of ten struck flints (consisting of flakes and a blade) and a single piece of burnt flint were recovered. The direction of deposition of this fill was from the west and its composition indicated it was likely derived from organic-rich activity-related refuse/material (*Type III*). It is unclear as to whether this deposit had either been dumped in a single episode or had gradually accumulated. Above (578), a small deposit of pale brownish yellow silty sand with no inclusions was present, fill (579), which is thought to represent a distinct deposit of wind-blown sand (*Type V*).

As was seen across the vast majority of the enclosure, the final tertiary fills in this part of the site were represented by *Type IV* deposits comprised of friable dark greyish brown silty sands with moderate small sub-angular flints (contexts (580; 582; 591; 596)). An assemblage of twenty-six struck flint flakes, a flint blade and a single sherd of a Middle Bronze Age urn which had fingernail impressions under the rim was collected from these fills.

Area B: South

Seven slots were excavated along the southern element of the enclosure in Area B. Four slots were excavated into the inner enclosure ditch (contexts (633=662=756=776)) and three slots were excavated into the outer enclosure ditch (contexts (631=642=676)) (Figure 8). The superficial geology across this part of the site is the same as in Area A, which was reflected in the make-up of the archaeological deposits recorded. The primary fills in this part of the site mostly consisted of *Type II* deposits made up of pale yellow clays which contained an abundance of pellet-sized (small, rounded and under 1 cm) chalk inclusions (deposits (634; 648; 677; 678; 679; 759; 760; 777; 779)), most likely representative of bank slump, from which only animal bone was recovered. A number of *Type I* deposits were also recorded amongst the primary fills across this part of the enclosure, and these comprised pale to mid-yellowish-brown silts with occasional small sub-angular flint inclusions (deposits (637; 638; 643; 644; 645; 646; 663; 778)). Four struck flint flakes were recovered from these fills, along with a small assemblage of animal bone. In a single instance, a deposit of wind-blown pale brownish yellow silty sand with no inclusions, fill (761) (*Type V*), was present above deposit (760) in slot (756). This was similar to deposit (579) recorded in the east of Area B.

Due to the slump direction of the *Type II* deposits of redeposited natural in this area of the enclosure ditch, it is very likely that a bank was present in-between the two ditches that made up the double-ditched enclosure in this part of the site.

The secondary deposits infilling the enclosure in the south of Area B were dominated by *Type III* deposits consisting of dark greyish brown sandy silts with moderate quantities of charcoal flecks of alder/hazel and *Prunus* species, and occasional small sub-angular flint inclusions (deposits (632; 647; 649; 762; 780)). From these fills, an assemblage of four struck flint flakes, burnt flint and animal bone was recovered. Despite the lack of pottery retrieved from these deposits, they can be taken as a good indication that anthropogenic activities occurred inside of the enclosure in Area B. Further sterile, washed-in, water-lain silts (635; 763), of a mid-orangey brown hue were also recorded among the secondary fills across this part of the enclosure (*Type I*). A small assemblage of animal bone was collected from deposit (635).

As was recorded across the rest of the enclosure, the tertiary deposits in the two ditches investigated in this part of the site consisted of mid-brown sandy silts with frequent small sub-angular flint inclusions (deposits (636; 639; 650; 651; 664; 680; 681; 764; 765; 781; 782)). These *Type IV* deposits most likely formed through the gradual accumulation of surrounding soils into the ditches, where they have subsequently leeched. An assemblage of twelve sherds of pottery (of Middle Bronze Age (59 g), Late Bronze Age (5 g) and earlier Iron Age (14 g) date), fifty-four struck flint flakes, a flint backed knife, burnt flint and animal bone was recovered from these deposits, along with a number of intrusive metal finds including three iron nails and a post-medieval copper-alloy button.

Area B: South-West

A total of three slots were excavated into the south-west corner of the enclosure. Here, the outer circuit had a break on the corner, resulting in two terminal ends, of which both were excavated (contexts (665; 671)). A slot was also excavated on the corner of the inner enclosure ditch (context (704)). The primary fills recorded on the outer enclosure ditches were represented by deposits made up of pale yellowish–greyish clays which contained an abundance of pellet-sized chalk inclusions, from which no finds were recovered (deposits (666; 672; 691; 692; 727)). These *Type II* deposits are most likely derived from the erosion of the bank and slumping of the up-cast deposits shortly after the ditches were dug. The primary fills of the inner enclosure ditch were not reached in this location due to the inflow of ground water, with the water table being reached 0.9 m below the stripped horizon at the time of the excavation.

Above these deposits, a complex of secondary silts was encountered (contexts (667; 673; 716)). These deposits consisted of mid-yellowish-brown clay-rich silts with occasional small sub-rounded chalk and sub-angular flint inclusions, from which no finds were collected. Such fills can broadly be categorised as *Type I* deposits, which likely formed through the accumulation of washed-in and water-lain fine silts from the immediate environment around the ditches. Stratigraphically, following their deposition in the outer enclosure ditches, organic-rich *Type III* fills (668; 674) were then deposited, which most likely derive from the gradual accumulation or single event dumping of anthropogenically formed material. These were composed of oak charcoal-rich mid-greyish brown silts with occasional small sub-angular flint inclusions, from which no finds were recovered. Their presence in the outer enclosure ditches (along with those identified in slots (631) and (642)) may point towards the occurrence of contemporary activity around the outside of the enclosure in the south of Area B. In the inner enclosure ditch, *Type III* deposits were not identified, and instead two deposits of pale to mid-brown silty clay with frequent pellet-sized rounded chalk inclusions were recorded (contexts (717; 718)), from which no finds were retrieved. These deposits are considered to represent slumping of the bank (*Type II*).

The final tertiary fills were consistent with those recorded across the rest of the enclosure, made up of greyish brown sandy silts with occasional small sub-angular flints and occasional charcoal flecks (*Type IV*) (deposits (669; 670; 675; 719=720)). Finds were only recovered from deposit (719) of the inner enclosure ditch and consisted of animal bone, three struck flint flakes, one piece of fired clay and five sherds of earlier Iron Age pottery.

From analysis of the *Type II* deposits of slumped redeposited natural in this area of the enclosure ditch, it seems probable that a bank was present in-between the two ditches that make up the double-ditched enclosure in this part of the site, due to the direction of mass wasting appearing to originate from this area (see deposits (666; 672; 692; 717; 718)). From the presence of deposit (727), it is possible that the bank may have also extended slightly around terminus (671).

Area B: West

Four slots were excavated across the western part of the enclosure ditches in Area B. Two slots were excavated into the inner enclosure ditch, (705=838), and two slots were excavated into the outer enclosure ditch, (682=753). The primary fills recorded in this part of the enclosure were represented by *Type II* mid-yellowish-brown silty clays, with frequent small chalk pellet and occasional small sub-angular flint inclusions, from which no finds were recovered (deposits (683; 684; 724)). From the direction of mass wasting, these deposits suggest the presence of a bank in-between the two ditches that made up the double-ditched enclosure in this part of the site.

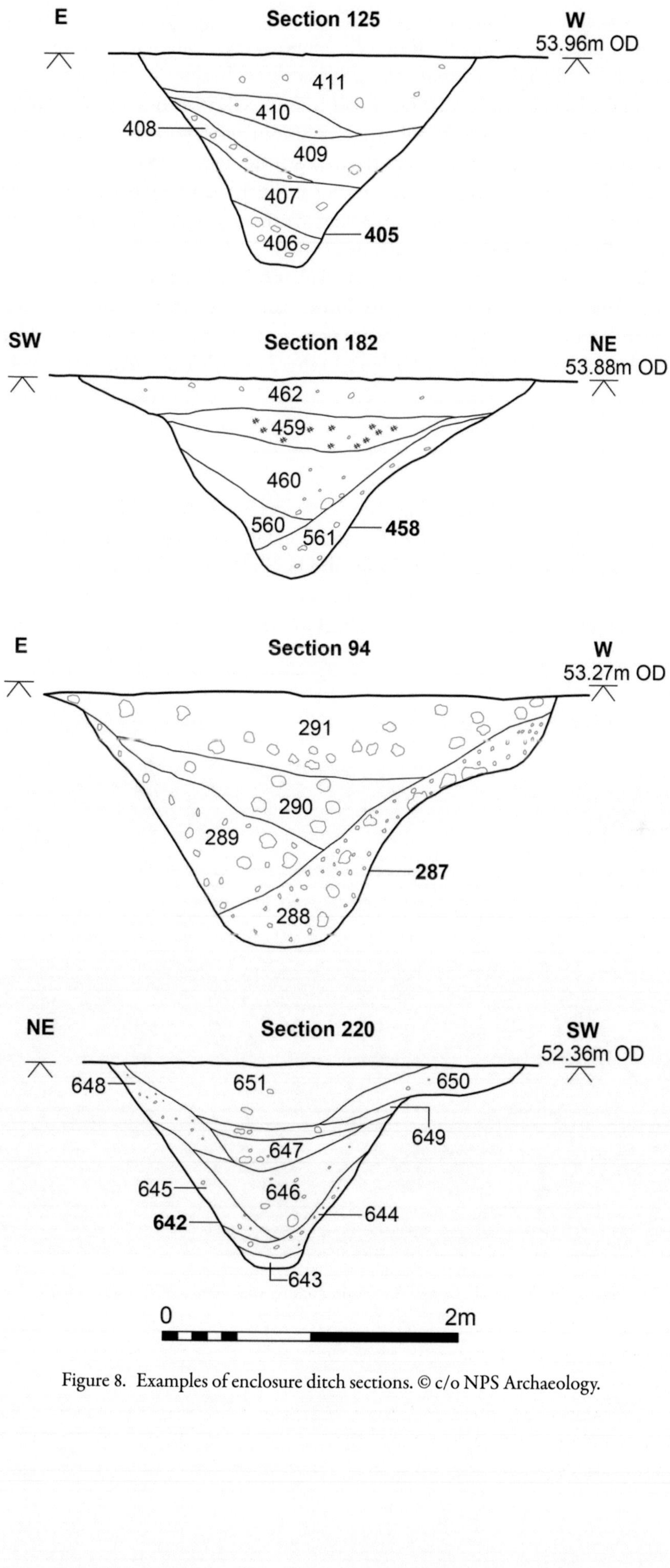

Figure 8. Examples of enclosure ditch sections. © c/o NPS Archaeology.

Secondary deposits were dominated by *Type I* mid-yellowish-brown clay-rich silts with occasional small sub-rounded chalk and sub-angular flint inclusions (contexts (685; 686; 687; 688; 689; 725; 754)), from which a small assemblage of animal bone was recovered. In slot (838), however, two deposits of redeposited natural were present consisting of pale to mid-yellowish-brown silty clay, with frequent small chalk pellets and occasional small sub-angular flint inclusions (*Type II*) (contexts (839; 840)). Four sherds of earlier Iron Age pottery were recovered from deposit (839), all of the same fabric. It is highly probable these deposits identify the clearance of bank material from the space between the two Middle Bronze Age ditches, prior to the casting of earlier Iron Age Linear 1 (also explaining the presence of the earlier Iron Age pottery in deposit (839)).

Tertiary fills were consistent with those recorded across the rest of the enclosure (*Type IV*); in the west of Area B they were made up of greyish brown sandy silts with occasional small sub-angular flint inclusions and rare charcoal flecks (deposits (690; 726; 755; 841)). A finds assemblage consisting of animal bone, burnt flint, a single flint blade and a sherd of earlier Iron Age pottery was recovered.

Burnt Mound Activity

An area of activity consisting of both features and deposits rich in burnt flint and charcoal was identified in the central southern part of Area A (Figures 7 and 9). This activity appears to have been concentrated in the space between the two enclosure ditches in this area. No upstanding 'burnt mound' deposits were confidently identified, but the features and deposits discussed below strongly suggest that such activities were conducted at the site. Full interpretation of these remains is presented in Chapter 6.

Plate 3. Enclosure ditch (361), 2 m scale, looking west. © c/o NPS Archaeology.

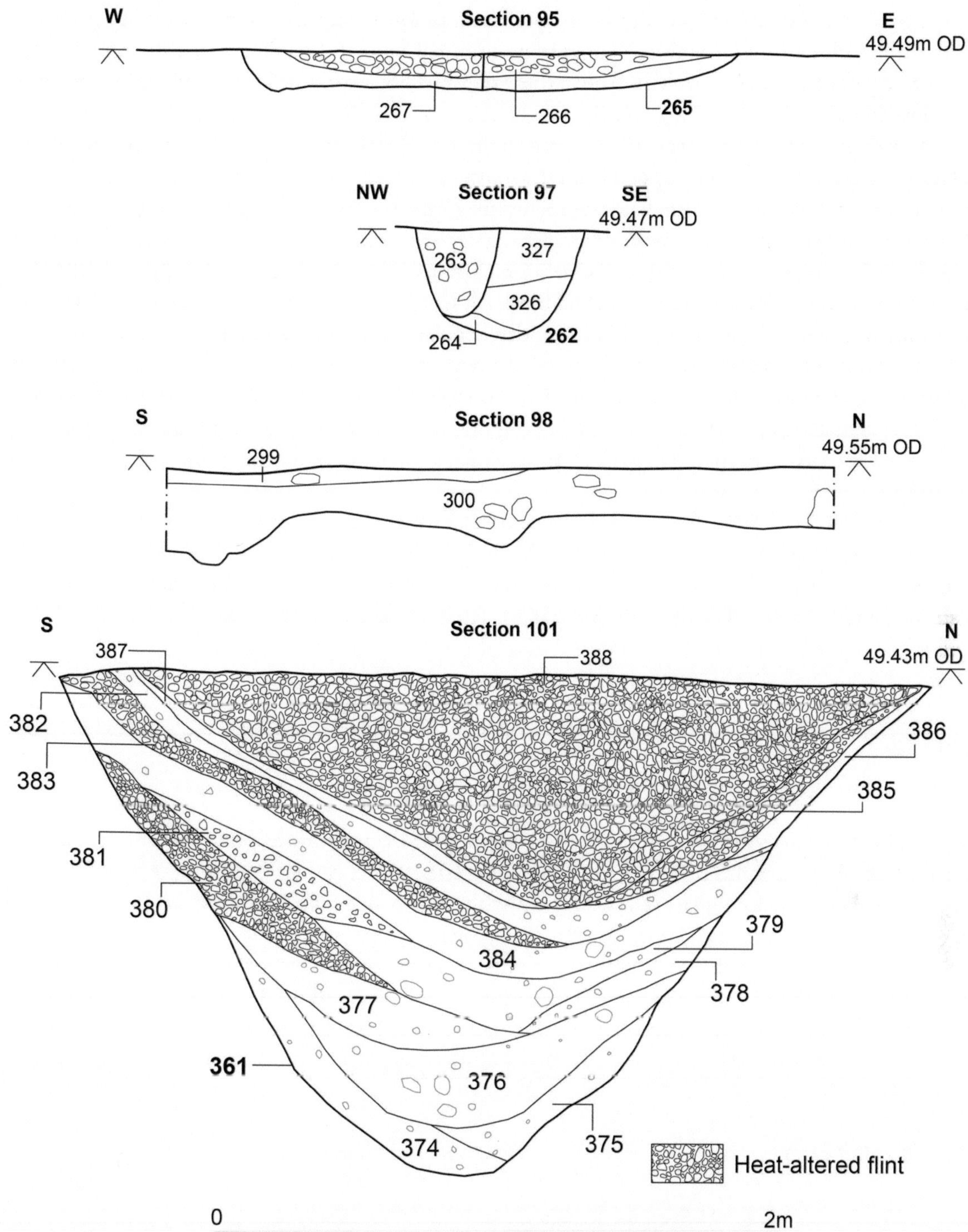

Figure 9. Sections of pit (265), post-hole (262), deposit (299) and ditch (361). © c/o NPS Archaeology.

Enclosure Ditch (361)

In the centre of the southern inner enclosure ditch in Area A, a unique deposit was identified in plan, which was subsequently investigated in slot (361) (Plate 3). The lower infilling sequence of this ditch was typical for the site and the deposits were similar to those seen across the rest of the enclosure, consisting of primary and secondary *Type I* and *Type II* deposits of mid-yellowish-brown clay-rich silts (376; 378) and pale-yellow silty clays with chalk pellet inclusions (374; 375; 377) respectively. From these fills, a finds assemblage consisting of animal bone, burnt flint and a sherd of Middle Bronze Age pottery was recovered. Above these deposits however, a unique stratigraphic sequence was identified.

Overlying deposit (377) was (380), a *c.*0.15 m thick deposit of small burnt flints in a charcoal and black silt matrix that had been tipped from the south (*Type VI*). This was overlain by (379), a deposit of pale-yellow silty clay with chalk pellet inclusions (*Type II*). Above this layer was (381), a further burnt flint and charcoal-rich deposit of greyish dark brown silt (*Type VI*). This was sealed by deposit (382), a pale-yellow silty clay with chalk pellet inclusions (*Type II*). Above this, (383), a further deposit of small burnt flints in a charcoal and black silt matrix which had been tipped from the south, was present (*Type VI*). In a repeated fashion, this deposit was in turn sealed by (384), a pale-yellow silty clay with chalk pellet inclusions (*Type II*). A break in this repeated sequence was represented by (386=387), which consisted of a pale yellowish-brown sterile water-lain silt with no inclusions (*Type I*). The two final deposits infilling the ditch, (385) and (388), were comprised of small burnt flints in a charcoal and black silt matrix (*Type VI*), from which a fragment of fired clay and animal bone was recovered.

Bulk samples of these deposits did not recover any plant macrofossils or microfaunal remains, but large quantities of charcoal were retrieved. Charcoal from deposit (376) returned a radiocarbon date of 1497–1305 cal. BC (Beta – 552729, 95.4 per cent) and charcoal from burnt flint deposit (388) returned a radiocarbon date of 1450–1291 cal. BC (Beta – 552728, 95.4 per cent).

Pit (265)

Approximately 3 m to the south of ditch (361) was an oval-shaped feature, pit (265). This feature had a length of 1.85 m, a width of 0.82 m and a depth of 0.14 m. The primary deposit (267), which consisted of a 0.04 m thick compact yellow silty clay with chalk pellet inclusions, represented redeposited natural used as a clay lining. Above the clay lining, inside the formed cavity, a 0.10 m thick deposit of small burnt flints in a charcoal and black silt matrix was present (contexts (266=296=297=298)), identical to the *Type VI* deposits seen in ditch (361). Two pieces of fired clay were recovered from this deposit, with no plant macrofossils or microfaunal remains recovered from the samples taken.

Post-Hole (262)

At 3.7 m to the north-west of pit (265), post-hole (262) with a sub circular diameter of 0.62 m was located. This feature had a 'U'-shaped profile with steep, sloping sides and a gentle sloping base. In the base of this feature, deposit (264), a 0.06 m thick deposit of friable mid-brown sandy silt with no inclusions was present, which possibly represents soil that fell into the feature shortly after it was cast. Above this deposit, a post had been placed, with two deposits of redeposited yellow clay natural used as post-packing (contexts (326; 327)). This post had been removed at a later date, and in this 0.3 m deep void, deposit (363), identical to the *Type VI* deposits seen in ditch (361) (consisting of small burnt flints in a charcoal and black silt matrix) was placed. The presence of this deposit in this feature along with its close juxtaposition to ditch (361) suggests that post-hole (262) represents a component feature of the burnt mound activity identified at the site.

Deposit (300)

A small amorphous roughly sub-oval in plan deposit of small burnt flints measuring approximately 1.7 m in diameter, and only *c.*0.02 m thick, was present *c.*7.5 m to the south-west of ditch slot (361). Little can be said in terms of the origin of deposit (300), but a possible suggestion is that it may relate to the previous location of a burnt flint mound. It is possible that some of the stockpiled burnt flints at the base of a potential mound moved through the soils onto the natural through bioturbation. However, an alternative interpretation for the origin of this deposit could be possible.

Post-Hole Groups

Seven groups of post-holes inside the enclosure in Area A are considered to be of Middle Bronze Age date. Although finds were not recovered in any abundance from these features, they have been assigned a Middle Bronze Age date due to their juxtaposition and spatial respect of the enclosure ditches and the 'burnt mound' activity. It is possible that some of these post-hole groups do not date to the Middle Bronze Age, and may instead date to the earlier Iron Age, when the enclosure was still in use. Consequently, the following section should be considered with this in mind.

Gate Structure 1

At the primary entrance/exit into the enclosure in Area A, defined by a termination in four of the ditches in the north-west of the site, an arrangement of eighteen post-holes was present (Figure 10). Post-holes in this group had a broad rectangular arrangement with a number of outliers and possible internal posts also present. With varying dimensions, most of these post-holes were sub-circular in plan with diameters ranging from 0.1 m to 0.46 m and depths from 0.06 m to 0.4 m. Profiles across the features in this group showed significant variation, but most had steep to vertical sides and flat or gently sloping bases. Post-holes in this group included features (401; 425; 427; 429; 431; 433; 435; 437; 439; 441; 443; 445; 447; 449; 451; 453; 853; 855). The relationship between intercutting post-holes (439) and (853), and (445) and (855) could not be established due to the identical nature of their fills. No post-pipes or post-packing deposits were identified across post-holes in this group.

The various deposits infilling the post-holes of this group all comprised similar mid-yellowish-brown sandy silts with occasional small sub-angular flints (deposits (402; 426; 428; 430; 432; 434; 436; 438; 440; 442; 444; 446; 448; 450; 452; 454; 854; 856)); with the exception of post-hole (437), which was filled with deposit (438), a mid-greyish brown sandy silt with occasional small sub-angular flint inclusions. No finds were recovered from any post-holes in this group.

Caution has been taken to not overinterpret the arrangement of the post-holes in this group and the possible structure, gate or fences that they formed. There is a significant likelihood post-holes in this group belong to different construction phases and/or relate to repairs and modifications of existing gates/structures/fences. A significant component of this group however was that two lines of posts appeared to block the areas between the enclosure ditches where banks were present, creating a possible passage way just over 2 m wide, perhaps to help guide people and animals in and out of the enclosure.

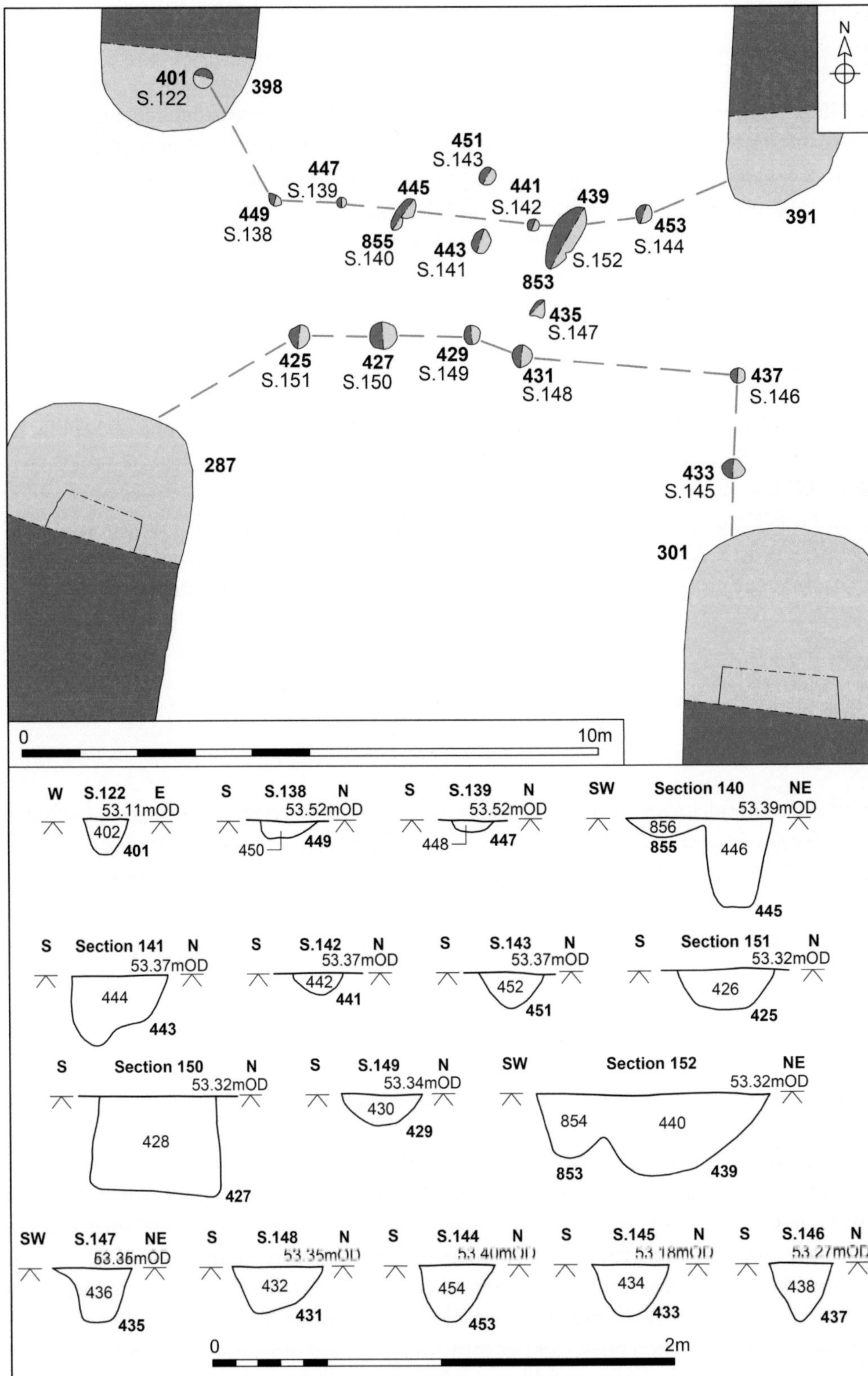

Figure 10. Gate Structure 1: Plan and sections. © c/o NPS Archaeology.

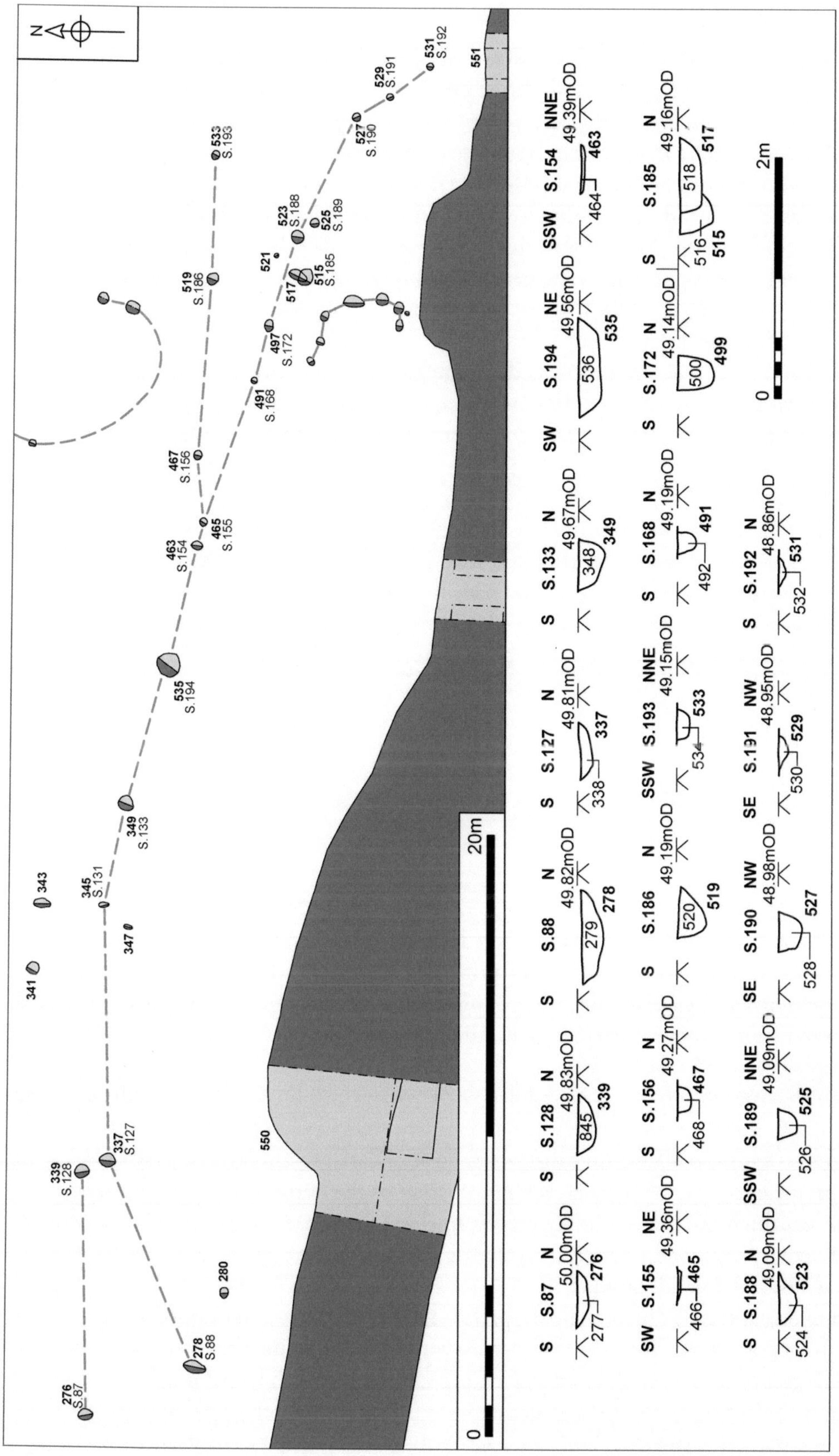

Figure 11. Fence-line 1: Plan and sections. © c/o NPS Archaeology.

Fence-Line 1

In the central southern part of Area A, a probable fence-line was identified (Figure 11). This group consisted of twenty-six post-holes of various forms (features (276; 278; 280; 337; 339; 341; 343; 345; 347; 349; 463; 465; 467; 491; 497; 515; 517; 519; 521; 523; 525; 527; 529; 531; 533; 535)). The diameter of the post-holes in this group ranged from 0.16 m to 0.83 m and their depths ranged from 0.03 m to 0.32 m. The possible fence-line had an approximate north-west-to-south-east alignment, curving slightly southwards at both ends; an arrangement reflecting the shape of the inner enclosure ditch in this area. The profiles of the post-holes in this group varied widely, but most frequently were of 'U'-shaped form with steep sloping sides and gently sloping bases. Two intercutting post-holes were identified, with (515) cut by (517). This suggests that repairs/modifications were made to the fence-line and also infers some longevity of use.

Three different deposit types were identified infilling the features in this group. Mid-dark greyish brown sandy silts with small sub-angular flint inclusions (deposits (277; 279; 281; 338; 342; 344; 346; 468; 492; 498; 518; 521; 524; 526; 528; 536)), mid-dark orange brown sandy silts with small sub-angular flint inclusions (deposits (845; 340; 348; 516; 530; 532; 534)) and dark grey clay-rich sandy silts (deposits (464; 466; 520)). No finds were recovered from any of the post-holes in this group.

It is likely the function of this fence-line was to block access, whether physically, visually or metaphysically, to the low-lying inner enclosure ditch area, where both the burnt mound activity took place and where the ditches and ground would have been most saturated at certain points of the year. This may have been to control animals, people or both. The fence-line was most likely of wattle construction and it is possible that supporting posts were employed between the post-holes in this group, which did not intrude significantly into the ground.

Structure 1

A group of eleven sub-circular post-holes were present to the north of Fence-line 1 (Figure 12). Seven of these features appeared to be arranged in a curved formation (features (469; 471; 473; 475; 479; 483; 485)), with four further probable associated post-holes located nearby (features (477; 481; 487; 489)). The dimensions of the features in this group varied, with diameters ranging from 0.24 m to 0.53 m and depths ranging from 0.08 m to 0.25 m. All the post-holes in this group had steeply sloping sides with gently sloping or flat bases.

Most of the features in this group were filled with mid-greyish brown clay-rich silts with rare small sub-angular flint inclusions (deposits (467; 472; 474; 476; 478; 482; 484; 486; 488; 490)), however post-hole (479) was filled with a distinct deposit of dark greyish brown silt with abundant small burnt flint inclusions (480). This fill was reminiscent of the burnt flint (*Type VI*) deposits recorded in enclosure ditch slot (361), possibly suggesting that Structure 1 had a related function to the 'burnt mound' activity identified, securely dated to the Middle Bronze Age. No finds were recovered from any of the post-holes in this group.

The form of Structure 1 is difficult to interpret; it may have comprised of only the curved formation made up of post-holes (469; 471; 473; 475; 479; 483; 485), or may have included the other post-hole in this group, (477; 481; 487; 489), forming something much larger and 'D'-shaped. It is possible that the posts in this part of the structure were not driven as deep, or that a greater thickness of subsoil was present in this area. Both these scenarios would prevent against the archaeological identification of potential post-holes – however this is just conjecture, and consequently the dashed

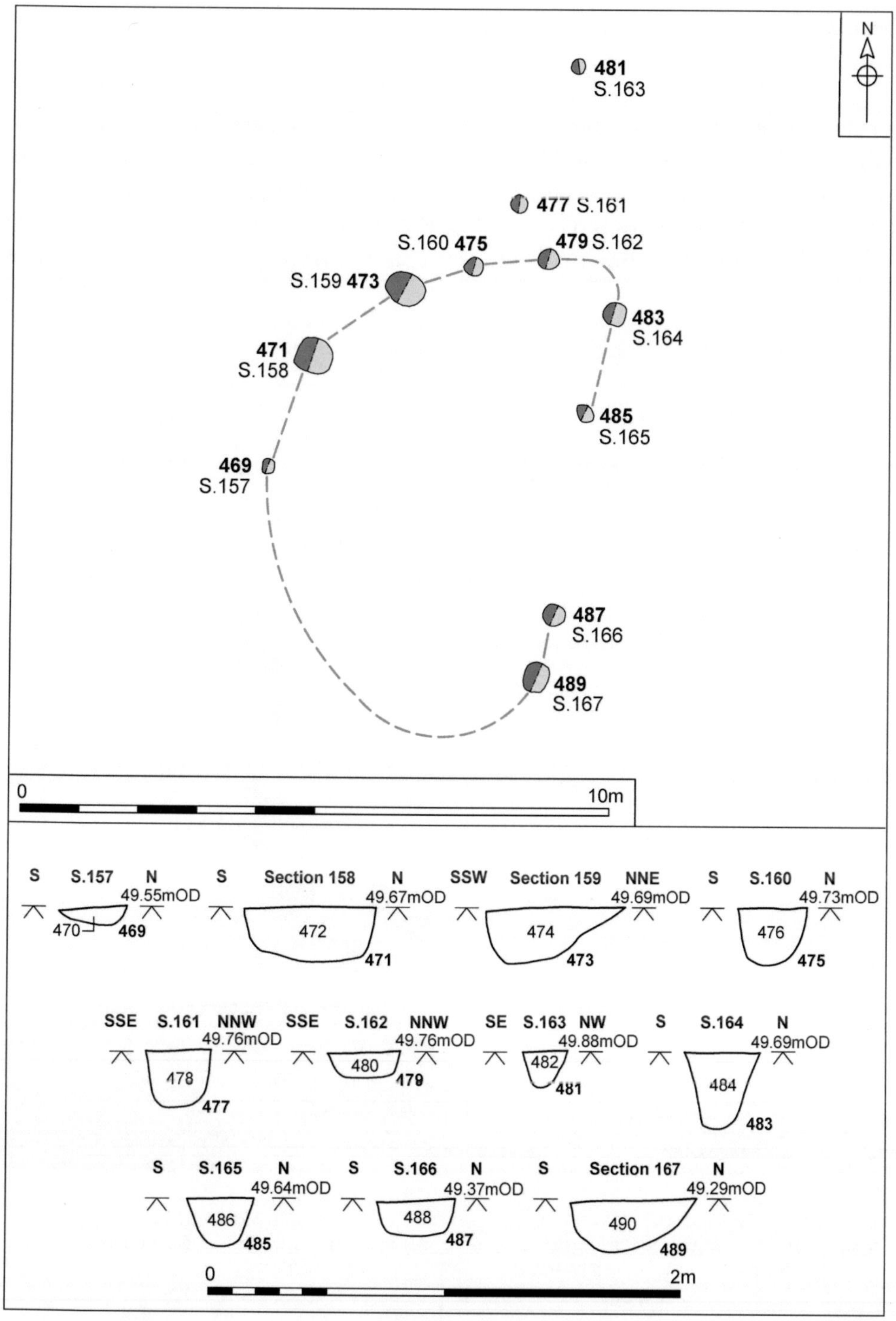

Figure 12. Structure 1: Plan and sections. © c/o NPS Archaeology.

line on Figure 12 should be considered as such. Conversely, if the former suggestion is correct, it is notable that the remains of Structure 1 and Structure 2 are of strikingly similar form. Based on the available evidence it is not possible to confidently assess the purpose/function of Structure 1, however the identification of deposit (480) may indicate it had a connection to the Middle Bronze Age 'burnt mound' activity.

Structure 2

A group of eight sub-circular post-holes was identified to the immediate north of the southern enclosure ditch in Area A (features (493; 495; 499; 501; 503; 505; 511; 513)) (Figure 13). Post-holes in this group were arranged in a 'C'-shaped curved formation, with diameters ranging from 0.19 m to 0.35 m and depths ranging from 0.13 m to 0.32 m. The features in this group had steep sloping sides with gently sloping or flat bases. All of the features in this group were filled with mid-greyish/

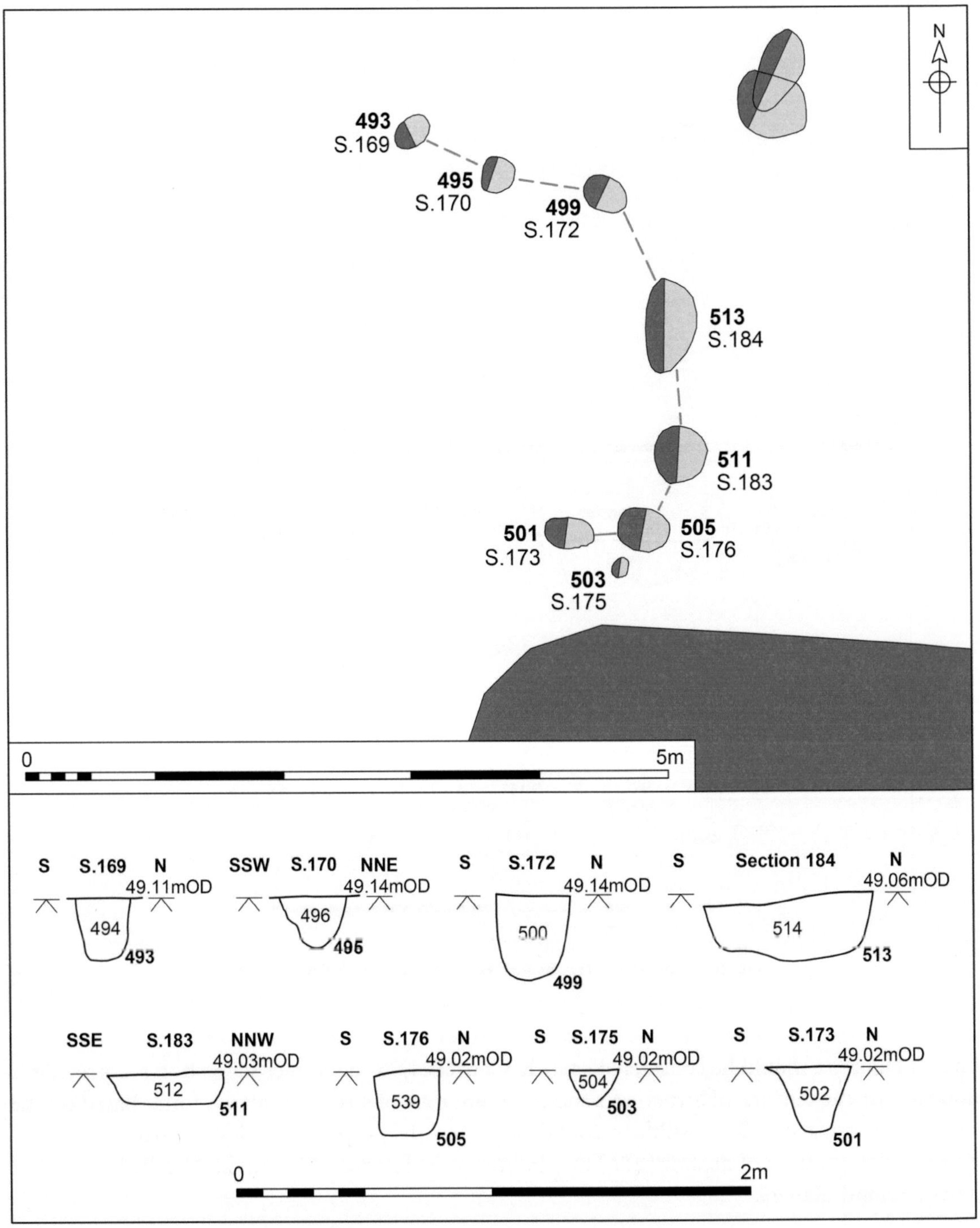

Figure 13. Structure 2: Plan and sections. © c/o NPS Archaeology.

orangey brown clay-rich silts with occasional small sub-angular flints (deposits (494; 496; 500; 502; 504; 506; 512; 514; 539)). No finds were recovered from any of the post-holes in this group. The curved formation of this post-hole group was similar to Structure 1. However, as with Structure 1, interpretation as to the function/purpose of this post-hole group is not clear.

Fence-Line 2

Fence-line 2 consisted of seventeen sub-circular in plan post-holes with near vertical sides and flat bases (features (154; 156; 158; 160; 162; 164; 166; 168; 170; 172; 174; 176; 178; 180; 182; 184; 186)) (Figure 14). The diameters of the features in this group varied from 0.22 m to *c.*0.45 m with depths varying from 0.1 m to 0.32 m. The post-holes in this group formed a slightly curving line of posts *c.*6.80 m long, with two further posts situated *c.*4.50 m to the north. The formation appeared to respect the north-western edge of the Middle Bronze Age inner enclosure ditch.

All of the features in this group were filled with deposits of dark greyish brown clay-rich silts with occasional burnt flint inclusions and charcoal flecks (deposits (155; 157; 159; 161; 163; 165; 167; 169; 171; 173; 175; 177; 179; 181; 183; 185; 187)). An assemblage of burnt flint weighing 1,455 g was recovered from the fills of the post-holes in this group. The presence of burnt flint suggests that the posts were not left to rot *in situ*, but were removed and the sockets subsequently infilled with burnt flint and charcoal-rich deposits (possibly derived from the burnt mound activity identified at the site).

The close-set nature of these post-holes was a key characteristic of this group, and might infer a wattle construction. The function of this fence-line is not clear, it may have blocked physical or visual access to either an activity area or part of the enclosure ditch. It may have alternatively acted as a windbreak of sorts, sheltering certain activities, or had a similar such purpose currently unidentified. It is possible that Fence-line 2 may be associated with pits (33) and (35).

Fence-Line/Structure 3

Fence-line/Structure 3 consisted of nineteen sub-circular post-holes with steep sloping sides and flat bases (features (216; 218; 220; 222; 224; 226; 227; 229; 231; 233; 235; 237; 239; 241; 243; 245; 247; 249; 251)) (Figure 15). The post-holes in this group had diameters of between 0.2 m and 0.65 m and depths ranging between 0.07 m and 0.35 m. The features in this group were all filled with deposits consisting of mid-dark orangey brown sandy silt with occasional small sub-angular flint inclusions (contexts (217; 219; 221; 223; 225; 227; 228; 230; 232; 234; 236; 238; 240; 242; 244; 246; 248; 250; 252)). One struck flint and ten pieces of burnt flint were recovered from the fills of the features in this group.

It is difficult to interpret the arrangement of the post-holes in this group, but there appears to have been a distinct linear arrangement of post-holes among the group in a north-west-to-south-east alignment. This post-line appears to lead from the Gate Structure 1 entrance in the north-west of the enclosure towards the 'burnt mound' activity area and where Structure 1, Structure 2, Fence-line 1 and Post-hole Group 2 were situated. It is considered most likely that this element of this group represents a fence-line, perhaps aligning a track leading from the primary entrance/exit and heading to what appears to have been the most significant activity area inside the enclosure. Alternatively, this group could represent the remains of a large structure oriented alongside a north-west-to-south-east track.

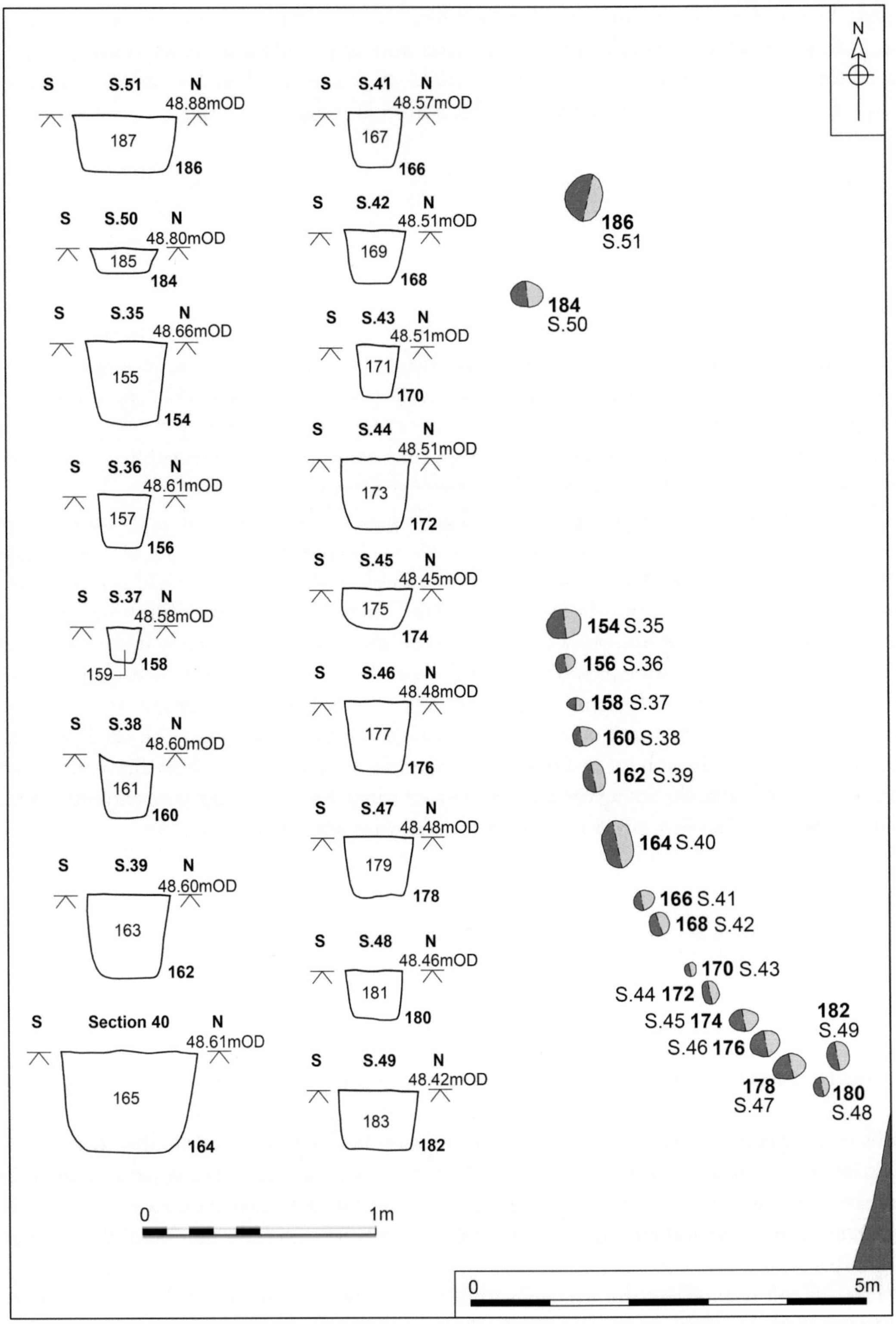

Figure 14. Fence-line 2: Plan and sections. © c/o NPS Archaeology.

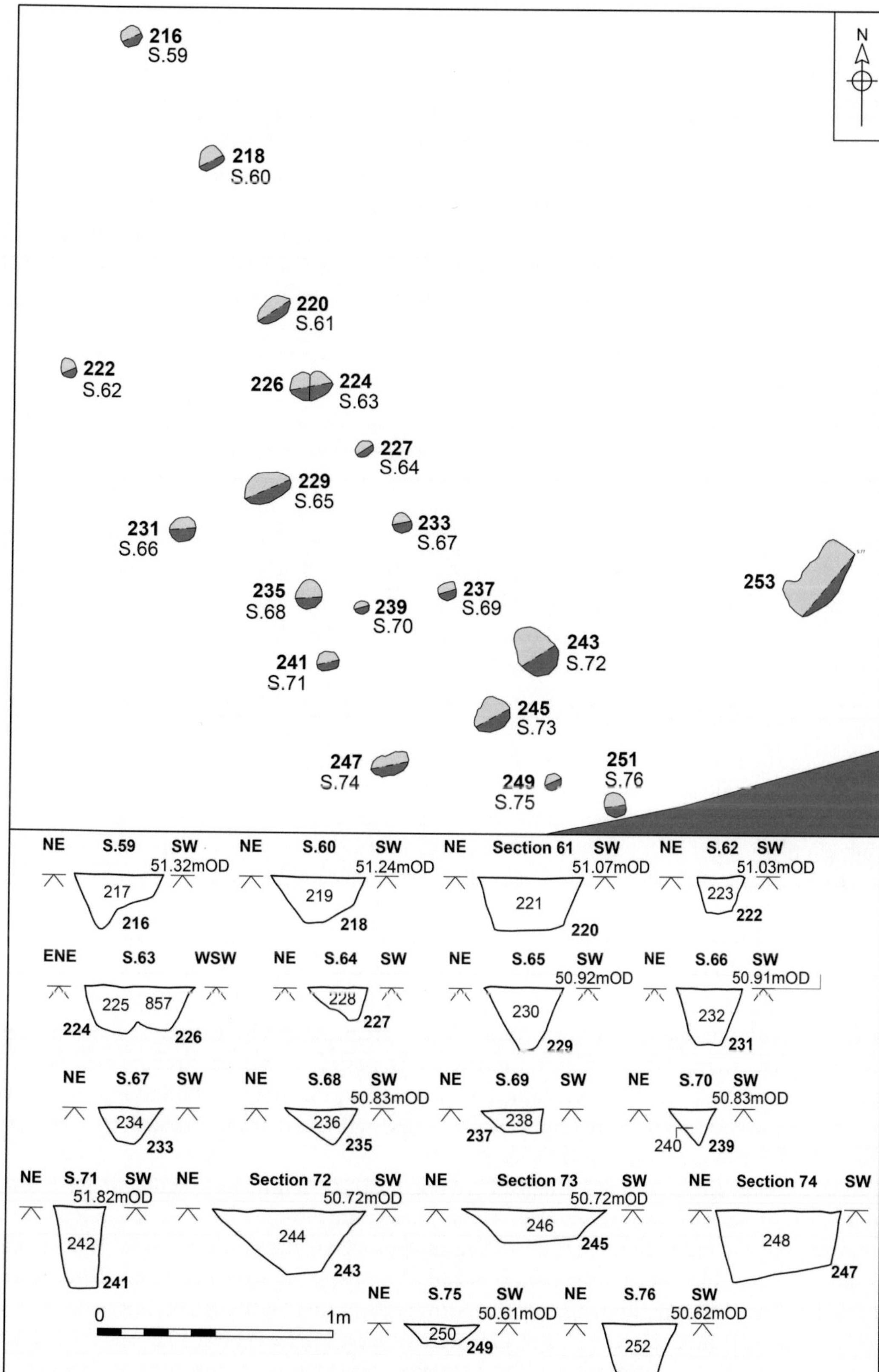

Figure 15. Fence-line/Structure 3: Plan and sections. © c/o NPS Archaeology.

Post-Hole Group 2

Four sub-circular in plan post-holes, (268; 270; 272; 274) forming a discrete group were identified to the south of Fence-line/Structure 3. Post-holes in this group had diameters of between 0.33 m and 0.52 m, depths ranging from 0.08 m to 0.24 m and bowl-shaped profiles with moderate sloping sides and bases. The post-holes in this group were filled with mid-orangey brown sandy silts with occasional small sub-angular flint inclusions (deposits (269; 271; 273; 275)). No finds were recovered from any features in this group. Post-hole Group 2 may represent a small fence-line, with a similar function to Fence-line 1, or may have equally had a different currently unidentified purpose.

Discrete Features

A number of discrete features were recorded and categorised as being of Middle Bronze Age date (Figure 7 and 16). The presence of burnt flint in their fills has been taken as an indication of their potential contemporaneity with the 'burnt mound' activity (although there is the possibility that some of these flints may be residual). Several discrete features have also been dated as Middle Bronze Age through their proximity to other features/groups of this period. It is important to emphasise that it is entirely plausible some features discussed below may date to the earlier Iron Age or other periods, as only indirect evidence has been used to attribute a period date. Discrete features (71), (658) and (660) have previously been discussed above.

Pits (33), (35) and post-hole (38) were located to the south-east of Area A, immediately to the west of Fence-line 2. Post-hole (38) had steep sloping sides, a moderate sloping base, measured 0.41 m deep and was filled with two fills; primary fill (37) of mid-orangey brown clay-rich silt with frequent small sub-angular flints, and secondary fill (36) consisted of a dark greyish brown sandy silt, with frequent small sub-angular flint inclusions. This feature was cut by both pit (33) and pit (35). Pit (33) was sub-circular in plan with a diameter of 0.98 m and a depth of 0.29 m. It had a moderate sloping asymmetrical profile with a flat base and was filled with two deposits; primary fill (32) consisted of a mid-orangey brown clay-rich silt, with small sub-angular flint inclusions, and secondary fill (31) consisted of dark brownish black sandy silt, with occasional small sub-angular flint inclusions and frequent charcoal flecks. Pit (35) had a distinct form and was sub-circular in plan with a diameter of 1.83 m, a depth of 0.13 m, moderate sloping sides and an undulating base. This feature was filled with deposit (34), a dark greyish brown sandy silt with frequent small sub-angular flint inclusions. From this deposit, 2,337 g of burnt flint was recovered. Deposit (34) was very similar to the fills in the Fence-line 2 post-holes, which also contained moderate quantities of burnt flint. It is possible that a direct association between these pits and Fence-line 2 existed, and that they collectively relate to the 'burnt mound' activity identified at the site.

Three close-set pits/post-holes, arranged in a north-east-to-south-west alignment, were also located in the south-east of Area A. Features (42) and (44) were sub-circular in plan, and feature (46) had an oval shape in plan. All three features had an approximate diameter of 0.9 m. With features (42) and (44), possible evidence for central stake-holes/post-pipes were identified. These features were filled with mid-orangey brown sandy silts with frequent small sub-angular flint inclusions (deposits (41; 43; 45)). Three pieces of burnt flint were collected from deposit (43), the fill of feature (44). It is not possible to confidently apply an accurate interpretation to these features, but they may represent part of a fence-line, a structure, a reinstated tethering post or similar.

Two discrete post-holes were present in the southern central part of Area A surrounded by Structure 1, Fence-line/Structure 3, Post-hole Group 2 and Fence-line 1. Their close juxtaposition to these post-hole groups suggests that they were broadly contemporary. Post-hole (351) was circular in plan with a diameter of 0.24 m and a depth of 0.04 m, with moderate sloping sides and an undulating base. It was filled with (350), a deposit of dark greyish brown clay-rich silt with occasional small sub-angular flint inclusions, from which no finds were recovered. This feature was either originally shallow or had undergone significant horizontal truncation. Approximately 1.1 m to the north-east, post-hole (353) was situated, which was sub-oval in plan with a length of 0.43 m, a width of 0.31 m and a depth of 0.28 m. This feature had a 'U'-shaped profile and was filled with (352), a dark greyish brown clay-rich silt with occasional small sub-angular flint inclusions.

Three moderate-sized sub-oval in plan pits were identified in Area A, inside the enclosure. Pit (370), which was truncated by post-medieval Ditch 5, had a length of 2.75 m, a width of 2.04 m and a depth of 0.6 m. This feature had steep sloping sides, a flat base and was filled with (371), a deposit of brown sandy silt with mixed orange and grey hues, that contained small sub-angular flint inclusions. A collection of burnt flint (422 g) was recovered from this deposit. This feature appeared to be paired with pit (372), which was located to its immediate west. Pit (372) had a length of 2.25 m, a width of 1.86 m and a depth of 0.4 m. It had an asymmetrical moderate sloping profile and was filled with (373), a deposit of brown sandy silt with mixed orange and grey hues, that contained small sub-angular flint inclusions, identical to deposit (372). Pit (253) was located to the immediate east of Fence-line/Structure 3. It had a length of 1.4 m, a width of 0.69 m and a depth of 0.22 m. It had steep sloping sides, an undulating base and was filled with (254), a deposit of dark greyish brown sandy silt with rare small sub-angular flint inclusions, from which 1,268 g of burnt flint were recovered. The

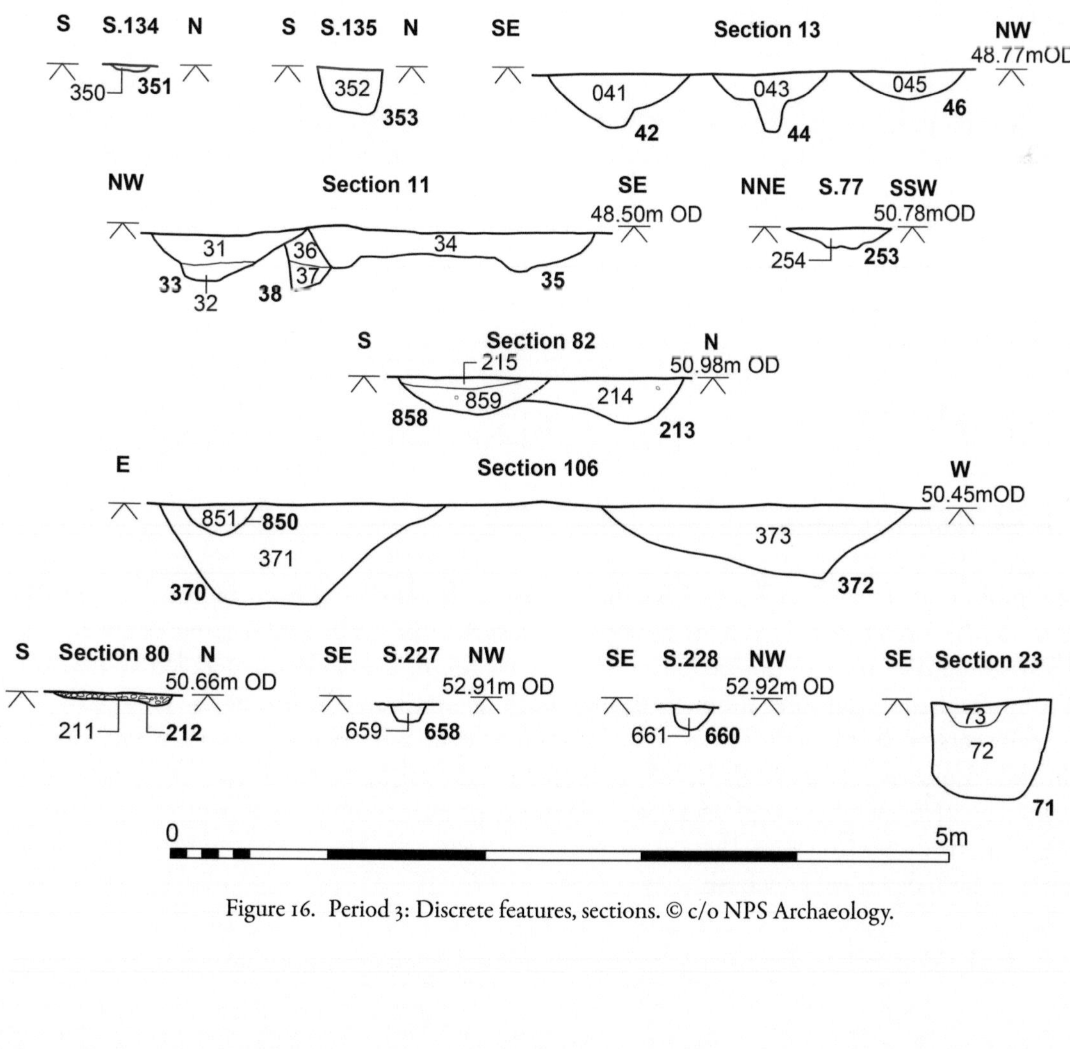

Figure 16. Period 3: Discrete features, sections. © c/o NPS Archaeology.

original function of these pits is unclear, but at the end of their useful lives, they all appear to have become infilled with similar materials.

Three further discrete pits were identified at the site and were situated on the outside of the enclosure in Area A. Pit (211) was distinct; it was neatly circular in plan with a diameter of 0.8 m. This feature had a flat bottom, but little more could be established about its profile, as it was extremely shallow (0.04 m deep). The shallow nature of this pit was either its original form or resulted from significant horizontal truncation. This pit was filled with (212), a deposit almost entirely composed of small angular fragments of burnt flint in a pale greyish brown sandy silt, with rare charcoal flecks. It is considered highly likely that this feature in some way relates to the 'burnt mound' activity identified at the site.

Pit (858) and pit (213) were located *c.*10.5 m to the north of pit (211). Pit (213) had an asymmetrical profile with a gentle sloping southern edge and a moderately sloping northern edge. It had a depth of 0.25 m and contained a single fill of pale orangey brown sandy silt with abundant small to medium sub-angular flint inclusions (deposit (214)). This feature was cut by pit (858), which had an oval shape in plan with a length of 1.06 m, a width of 0.71 m and a depth of 0.22 m. This feature had a symmetrical bowl-shaped profile and contained two fills; primary deposit (859), which consisted of a pale orangey brown sandy silt with abundant small sub-angular flint inclusions, and secondary deposit (215), that consisted of a dark greyish brown silt with occasional small sub-angular flint inclusions. From deposit (215), an assemblage of 246 g of burnt flint was collected.

Period 4: Late Bronze Age

A collection of fourteen sherds of Late Bronze Age (1200–700 BC) pottery was recovered from two discrete tertiary deposits infilling the Middle Bronze Age enclosure ditches. Deposit (545) from the inner enclosure ditch in the south-west of Area A contained an assemblage of five sherds of Late Bronze Age pottery, and deposit (782) from the inner enclosure ditch in the south of Area B contained nine sherds of Late Bronze Age pottery. Their presence in the upper sequence of the ditch, and the fact that neither deposits contained material culture dating from other periods indicates that they are unlikely to be finds of an intrusive or residual nature. These deposits most likely represent small-scale or transient use of the site between *c.*1200–700 BC.

Period 5: Earlier Iron Age

The Double-Ditched Enclosure

An assemblage of eighty Post Deverel-Rimbury pottery sherds dating from the earlier Iron Age (800–200 BC) was recovered from the upper fills across the Middle Bronze Age enclosure ditches. It is considered that many of the secondary and tertiary fills of the enclosure detailed above likely date from the earlier Iron Age; although this has been demonstrated to not be the case across the entire enclosure, with two of the tertiary (final) infilling deposits for example, radiocarbon dated to the Middle Bronze Age (deposits (361; 388; 458; 461/462)). Only a small number of deposits which contained notable assemblages of Post Deverel-Rimbury pottery could be confidently categorised as dating from the earlier Iron Age. These are deposits (291; 331; 396; 400; 410; 416; 548; 559) – all of which have been described previously.

The excavated evidence indicates that by the earlier Iron Age, the enclosure was partially filled, with some areas of the site already completely infilled. However, the earthworks of the Middle Bronze Age ditches and banks must have still been significantly open and visible in the landscape to allow for the reestablishment of activity at the site. Although it is clear ditches at the site filled up at different rates, in different parts of the enclosure and at different times, the precise mapping of deposition rates across the whole of the enclosure is severely limited by the scarcity of dateable finds.

Spatially, earlier Iron Age ceramics were recovered from across all areas of the enclosure, suggesting activity in this period was not confined to discrete parts of the site. The highest quantities of Post Deverel-Rimbury ceramics were recovered from the ditches around the north-east entrance/exist identified in Area A, indicating that it was most likely still in use during the earlier Iron Age. Significantly, no recutting of the Middle Bronze Age enclosure ditches appears to have occurred during this period.

Linear 1

The only modification to the enclosure that took place during the earlier Iron Age was the casting of Linear 1 in the west of Area B (contexts (697=701=751=831)). A 44.2 m length of this feature was present in Area B, continuing under the south facing baulk. It terminated outside of Area B, with the feature not continuing into Area A. It was situated between the inner and outer enclosure ditches for most of its length, slightly cutting the outer enclosure ditch (753) at its south-west end.

Linear 1 had a width of *c*.6.65 m and was orientated NNE-to-SSW, with an asymmetrical profile consisting of a gently sloping north-west side that had an indistinct break of slope (top) and a moderate sloping south-east side. On its south-east side, it had a depth of 0.6 m from the stripped

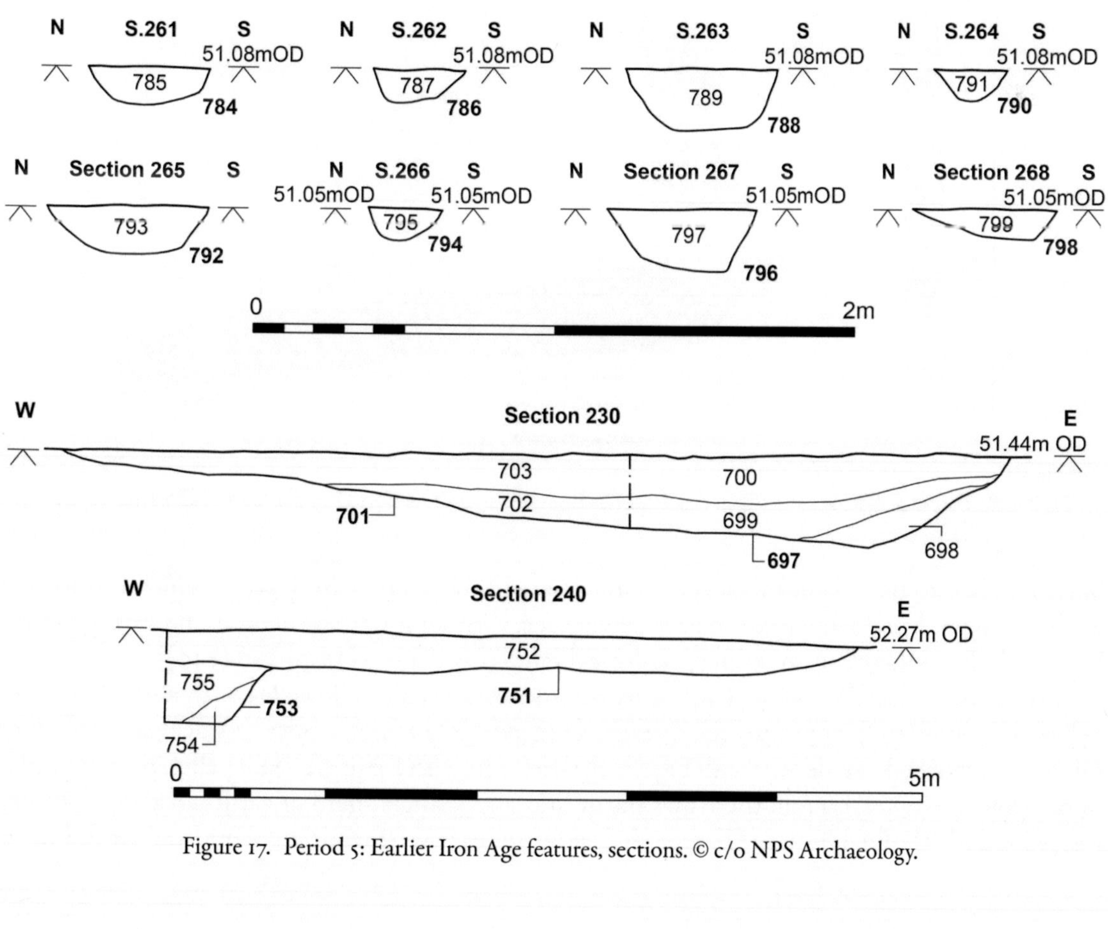

Figure 17. Period 5: Earlier Iron Age features, sections. © c/o NPS Archaeology.

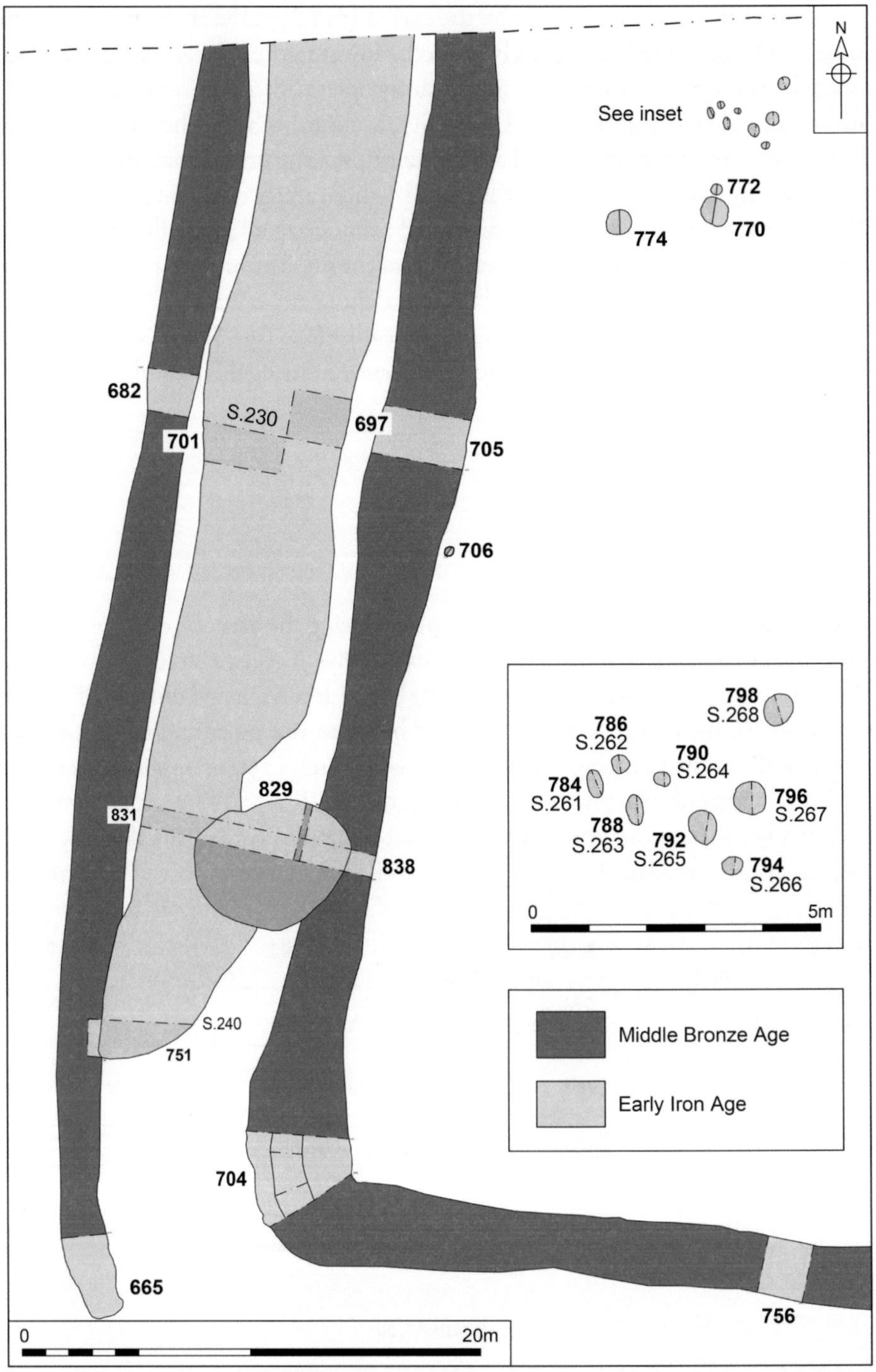

Figure 18. Period 5: Plan of earlier Iron Age features. Inset: detail of Post-hole Group 3. © c/o NPS Archaeology.

horizon. Prior to the casting of Linear 1, some clearance of the up-cast deposits which formed the Middle Bronze Age bank appears to have taken place, with some of these deposits appearing to have been backfilled into the surrounding enclosure ditches (e.g. deposits (839) and (840)).

The feature was filled with (698), a primary deposit of pale-yellow clay with abundant chalk pellet inclusions (698), appearing to represent a redeposited natural. Above this deposit, secondary fills were comprised of pale yellowish-brown clay-rich silts with frequent chalk pellet inclusions (deposits (699=702; 832)). From these fills, burnt flint and a single sherd of earlier Iron Age pottery was recovered. The final tertiary fills recorded across Linear 1 were made up of mid-yellowish-brown sandy silts, with occasional small sub-angular flints (deposits (700=703; 752; 833)). An assemblage of burnt flint, three worked flints (including a residually deposited Mesolithic adze) and three sherds of earlier Iron Age pottery was recovered from the tertiary fills. The make-up of the secondary and tertiary deposits suggests this feature was not backfilled and instead gradually infilled through the natural movement of soils.

The profile of Linear 1 was unique for the site, and this, combined with its short length, may suggest that its function was not as a ditch in the conventional sense; with the possibility that it may have represented a linear quarry hollow, a drainage sump to collect water (in what may have been the wettest part of the site) or had a similar function. However, its irregular and shallow form may alternatively be explained through an attempt to mimic the partially infilled enclosure ditches at the site in this period, and perhaps it does in fact represent an addition and enhancement of the Middle Bronze Age earthworks.

Post-Hole Group 3

A cluster of eleven post-holes was present to the north-west of Area B inside the enclosure ditch. This group consisted of features (770; 772; 774; 784; 786; 788; 790; 792; 794; 796; 798). Dating this group is tentative and premised on the recovery of an Iron Age loom weight from post-hole (788), along with the proximity of this group to earlier Iron Age Linear 1. There was some degree of variation in the form of features in this group, which were sub-circular in plan with diameters ranging from 0.24 m to 1.09 m and depths ranging from 0.18 m to 0.4 m. Based on the slightly larger diameter of features (770) and (774), it is possible that they may alternatively represent pits, although the depths, profiles and fills of these two features did not vary from the others in this group. The profiles of the features in this group were relatively symmetrical, with steeply sloping sides and gently sloping bases. They were all filled with similar friable dark orangish brown sandy silts with moderate small sub-angular flint inclusions, further suggesting they represent a coherent group (deposits (771; 773; 775; 785; 787; 789; 791; 793; 795; 797; 799)).

The original function of this post-hole group is unclear, but it was most likely structural in nature, as they are clustered, as opposed to being in a linear arrangement. Although a loom weight was recovered from the fill of one of the post-holes in this group, this cannot be taken as direct evidence they formed a loom or a domestic building.

Period 6: Romano-British

Two Romano-British (AD 43–410) artefacts were recovered from the site during the excavation. A single steelyard weight was recovered from probable marl pit (507=509) in the south-west of Area A. This item is relatively worn and is considered to have been residually deposited within a later feature. A sherd of possible Romano-British greyware pottery was also recovered from topsoil (722). During the trial trenching, a small assemblage of Romano-British pottery was collected from the Middle Bronze Age enclosure ditches (Peachey 2016a). These finds were probably deposited into partially infilled ditches, and possibly relate to the Early Romano-British activity identified *c.*200 m south-west of the site (Sleap and Savage 2016).

Period 7: Medieval

As with the Romano-British period, medieval (AD 1066–1540) activity at the site appears to have been of a small-scale or transient in nature. Five sherds of medieval pottery were recovered from the site. Two sherds came from topsoil (01), one was recovered as a residually deposited sherd from pit (258) and two further sherds were recovered as intrusive artefacts from the upper fills of ditches (287) and (652). A residual fragment of fourteenth–fifteenth-century Flemish floor tile and a fragment of medieval horseshoe were also recovered from the fills of post-medieval Ditch 3 and Ditch 2 respectively.

Period 8: Post-Medieval

Introduction

Five ditches, a probable hedge-line and one pit identified during the excavations dated to the post-medieval period (AD 1540–1900). Two of these ditches correspond with field boundaries depicted on the 1840 Swaffham Tithe Map, which are subsequently absent on the 1879–1886 First Edition Ordinance Survey Map (Ditch 3 and Ditch 4). The date at which these ditches were cast is not clear, but with no evidence of recutting, an eighteenth–early nineteenth-century date is considered most likely. The date of the other ditches identified is less certain, but again as discussed below, an eighteenth/nineteenth-century date is feasible.

Field Boundaries

Ditch 1, with a north-east-to-south-west alignment, was present to the north of Area A for a length of 26.45 m. Two slots excavated across its length revealed this ditch had an average width of 1.05 m and depth of 0.36 m. It had a concave profile and contained two fills. These fills contained a sherd of glazed red earthenware pottery, ceramic building material and iron finds including nails and a buckle, all suggesting a broad post-medieval date. This boundary is visible on an aerial photograph dating from 1946, indicating that the ditch was still likely in use into the mid-twentieth century. Its absence on later images also suggests it was backfilled around this point too. It is possible that the upper fill excavated represents topsoil used to infill the ditch, with the post-medieval finds within it originating from the surrounding plough soil.

Ditch 2 was also part of the same north-east-to-south-west aligned strip field arrangement, and was present in Area A. This ditch terminated in the north-east of the site and extended south-west for a length of 142.46 m within the excavation area and is presumed to have continued further beyond this. A total of ten sections were excavated along the length of this ditch to determine its form and recover artefactual evidence. The ditch had moderate sloping sides and base, with an average depth of 0.47 m and average width of 0.9 m. It was filled with a deposit of sterile clay-rich silt, similar in composition to the surrounding natural geology, suggesting it may have been infilled with previously up-cast material, perhaps from a bank. A finds assemblage consisting of a fragment of medieval horseshoe, an iron nail, a prehistoric flint side scraper, animal bone and post-medieval ceramic building material was recovered from this ditch. The horseshoe and flint scraper clearly represent residual finds deposited in the feature later. Unlike Ditch 1 or Ditch 3, this ditch is not present on any historic maps or aerial photographs, and thus likely went out of use prior to the mid-nineteenth century.

Ditch 3 shared the same alignment as Ditch 1 and Ditch 2 and had a probable hedge-line on its south-east side (Hedge-line 1). It extended for 180.9 m across Area A, continuing beyond the limits of the excavation in both directions. The ditch had an average depth of 0.65 m and width of 1.59 m. The infilling sequence of this ditch was distinctive amongst the post-medieval features; a mid-orangey brown sandy silt was present across the base of the feature, above which a field drain of nineteenth-century date had been laid in the partially infilled ditch. It is considered probable that although there was a need to backfill the ditch and combine the two fields into one (as is seen on the 1879–1886 First Edition Ordinance Survey Map), there was also a need to retain drainage of the heavy clay soils. From analysis of historical maps, this episode of backfilling likely occurred between 1868 and 1886. Above the field drain, a deposit of mid-greyish brown sandy silt was present which has been interpreted as material used to deliberately backfill the ditch following the laying of the field drain. The only finds recovered below the horizon of the field drain were a single nail and a fragment of animal bone. Above the field drain, broken fragments of ceramic drainpipe, a residual fragment of medieval Flemish floor tile, a single iron nail, a fragmented oyster shell and a sherd of post-medieval iron-glazed blackware were recovered. This ditch was recognised in Trench 4 of the evaluation, but not in Trench 3A (Smith 2012; Bull and Smith 2016) or by the geophysical survey, suggesting Ditch 3 terminated at the extant western field boundary in Area A.

As previously mentioned, a shallow linear feature interpreted as the footprint of a hedge-line was present along the south-east side of Ditch 3. Hedge-line 1 measured 0.05 m deep and 2.2 m wide, but only a *c.*50 m stretch was present. It is likely that this feature, if correctly classified as a hedge-line, extended along the full length of Ditch 3, with the element visible on site representing an area of intrusive rooting or a part of a bedding/removal trench. This feature had a moderate sloping south-east side, a flat base and was filled with a pale greyish brown sandy silt with abundant chalk flecks. Two sherds of post-medieval bottle glass were recovered from its fill. It is most likely that the removal of Hedge-line 1 also occurred between 1868 and 1886.

A further ditch of post medieval date extended across Area A and Area B. Ditch 4 was present for a total length of 107.8 m within the excavation areas. It continued beyond Area B to the south-east and terminated at the extant pond in Area A. It had an average width of 0.7 m, and depth of 0.35 m, with moderate sloping sides and a gentle sloping base. A single deposit of mid-brown clay-rich silt filled this ditch, which contained a finds assemblage of a residual struck flint flake, animal bone and post-medieval dated artefacts of an iron ferrule, an iron vessel, an iron hinge and ceramic building material. From analysis of the 1840 Tithe Map, this boundary is shown as connecting to/respecting Ditch 3, where it terminates. This boundary is shown as having been removed between 1868 and 1886, most likely in the same episode of boundary remodelling when Ditch 3 was backfilled and Hedge-line 1 removed.

Other Features

Ditch 5 was of north-west-to-south-east orientation and present in both Area A and Area B. It extended from the south of Area B, into the centre of the extant pond in Area A, present for a total length of 90.73 m within the excavation areas. The ditch had an asymmetrical profile with a concave western sloping side and a moderately sloping eastern side. Across its length it had an average width of 0.42 m and average depth of 0.15 m and was filled with a uniform mid-brown clay-rich silt. Although no artefacts were retrieved from this feature, it appears to have had a direct association with the extant pond in Area A, which dates from at least the late nineteenth-century. Despite the pond still being open, as Ditch 5 is sealed by subsoil (02=723), it is unlikely to be of modern date. This ditch probably served as a small drainage feature for the Area B field, towards the pond.

A single pit of post-medieval date was partially exposed and recorded in Area A. Pit (258) was an amorphous, sub-circular feature in plan with an exposed diameter of 3 m. It was 0.3 m deep with moderate sloping sides and a flat base. It was filled with a mid-yellowish-brown slightly sandy silt from which an iron nail, a sherd of residual medieval pottery, post-medieval ceramic building material and pottery of seventeenth–eighteenth-century date were recovered. The function of this pit remains unclear but may relate to an area of localised marl extraction. Its close proximity to undated feature (507=509) may suggest some relationship in function and date between these features, however this remains speculative.

Soil Overburden

As previously indicated, all post-medieval features at the site (including those of suspected late nineteenth-century date) were truncated by a pale orangey brown clay-rich sandy silt plough soil (02=723) which contained occasional sub-angular small flint inclusions. Consequently, it is likely the formation of this deposit is fairly recent. No dateable artefacts were recovered from deposit (02=723), with only a small assemblage of animal bone collected.

A number of post-medieval finds were recovered from topsoil (01=18=722) including an iron stake, an iron strip, four sherds of pottery, eight fragments of ceramic building material and two fragments of clay tobacco pipe. These finds may have been ploughed out of post-medieval features, result from casual surface loss and discard, or from night soiling.

Period X: Undated

A total of thirty-seven features of archaeological origin were not assigned a date, including a ditch, two large pits, ten small pits, twenty-three post-holes and a possible tree throw.

Area A

A large sub-circular pit, cutting the Middle Bronze Age enclosure ditches was present in the south-west corner of Area A. Pit (507=509) had a diameter of *c.*29.6 m and a depth of 1.04 m (ascertained through hand-auger). It had a gently sloping profile and was filled with a pale orangey brown sandy silt, from which a Roman steelyard weight was recovered (probably residual). It is tempting to connect this large feature to the extant pond in the same location, and perhaps they have a common origin. It is unclear as to whether (507=509) originally represents a marl pit or the early incarnation of a pond across what is the lowest lying and wettest part of the site; it is likely that throughout its history it functioned as both. This feature was probably of post-medieval date, however this cannot be proved beyond reasonable doubt.

Ten small isolated pits with diameters ranging from 1.07 m to 1.61 m, and depths ranging from 0.28 m to 0.7 m were identified in Area A (features (26; 28; 40; 202; 204; 208; 364; 366; 368; 417)). These pits were mostly filled with pale to mid-orangey brown sandy silts with occasional small sub-angular flint inclusions and contained no finds.

Fourteen probable post-holes with diameters ranging from 0.38 m to 0.72 m, and depths ranging from 0.02 m to 0.24 m were identified in Area A (features (49; 206; 304; 306; 308; 310; 312; 314; 316; 318; 320; 322; 324; 335)). These post-holes were mostly filled with pale to mid-orangey brown sandy silts with occasional small sub-angular flint inclusions and contained no finds.

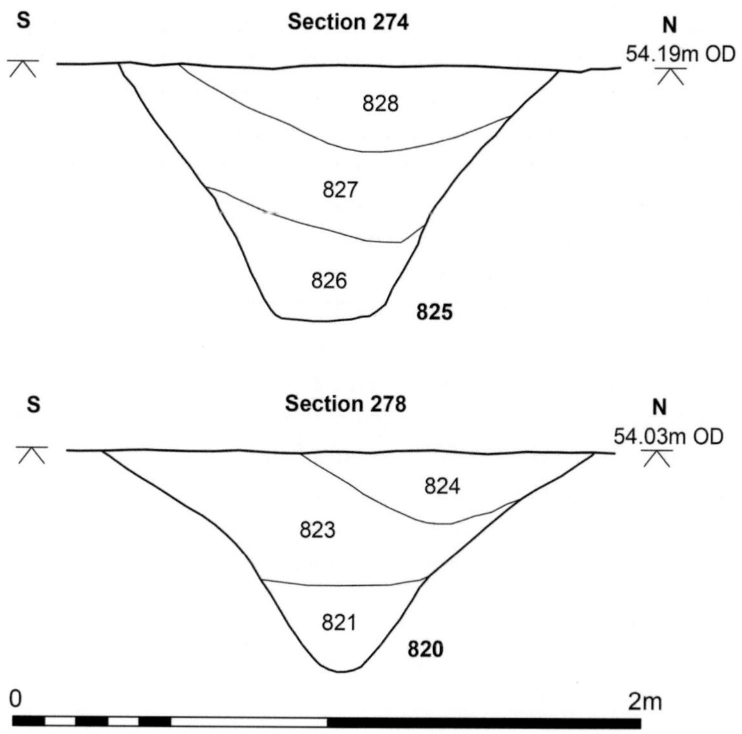

Figure 19. Ditch 6: Selected sections. © c/o NPS Archaeology.

Area B

Aligned WNW-to-ESE, Ditch 6 was present to the south-west of Area B for a length of 99.17 m (contexts (652=800=804=808=812=816=820=825=834)) (Figures 4 and 19). It had a symmetrical sloping profile with steep sloping sides and a flat base. Across its length it had an average width of 1.13 m and average depth of 0.75 m. Three distinct fills were identified across its length. Across the base of this feature, primary fills of pale brownish yellow clay-rich silts with rare chalk inclusions were identified (deposits (653; 801; 805; 809; 813; 817; 821; 826)). Above these fills, secondary deposits of mid-yellowish-brown sandy silts with rare sub-angular flint inclusions were recorded (deposits (655; 802; 806; 810; 814; 818; 823; 827)). The final tertiary fills in this ditch were composed of mid-brownish yellow sandy silts with occasional clods of redeposited chalky clay natural (deposits (657; 803; 807; 811; 815; 819; 824; 828; 835)). The presence of clods of redeposited clay natural in the tertiary fill may indicate that this feature was backfilled with bank material. The only finds recovered from this ditch consisted of a single small sherd of medieval pottery (3 g), a struck flint flake and a fused modern iron nut and bolt.

This feature has been categorised as undated due to the conflicting strands of evidence which could potentially be indicative of its date. Significantly, the ditch is on an identical alignment to the southern length of the enclosure in Area B, which could be taken as a strong suggestion that this feature may be of Middle Bronze Age/earlier Iron Age date. However, the presence of medieval pottery and a modern nut and bolt casts doubt over this, although equally these finds could be intrusive as they were recovered from the very top of the tertiary fill.

To the west of Area B, pit (829=836) was identified cutting earlier Iron Age Linear 1 and the Middle Bronze Age inner enclosure ditch (838). This feature was sub-circular in plan with a diameter of 6.7 m. It had a stepped profile which gently sloped from the top of the natural, with its sides subsequently becoming vertical. This feature was only excavated to a depth of 0.93 m due to the incursion of groundwater, with the features total depth established by hand-auger as *c.*1.8 m. It was filled with (830=837), a mid-orangey brown sandy silt with moderate small sub-angular flint and chalk pellet inclusions. Three small residual struck flint blades were recovered from this fill. The function of this feature is not clear. It could represent a marl extraction pit, which are abundant in the landscape, and may have had a secondary function as a drainage pond (as is potentially seen to the north). Alternatively, it may have been a rudimentary well or had another unidentified purpose.

An amorphous feature (640) present to the south of Area B continued beyond the edge of excavation. This feature had a width of 2.2 m, a depth of 0.31 m and an undulating irregular profile. It was filled with deposit (641), a mid-orangey brown clay-rich silt with rare small sub-angular flint inclusions, from which no finds were recovered. The irregular form and lack of finds recovered from this feature make it difficult to interpret, but it is possible that this feature could represent part of a tree throw.

Nine undated post-holes were identified in Area B (features (574; 612; 614; 616; 618; 706; 708; 710; 712; 714)). Most were discrete in nature with few indications that they had a common function, with the exception of post-holes (708), (710) and (712), which were all of a similar size (*c.*0.26 m diameter and *c.*0.20 m deep) and possibly relate to a fence-line, structure or similar. The undated post-holes in Area B were filled with deposits of pale to mid-yellowish and greyish brown silts, with occasional small sub-angular flint inclusions, from which no finds were recovered.

CHAPTER 3

Archaeological Finds

The artefacts retrieved from the excavation consisted of prehistoric pottery, a small assemblage of post-prehistoric pottery, ceramic building material, struck flint, burnt flint, non-flint stone, metal objects, fired clay, fragments of clay tobacco pipe and fragments of glass. In this chapter, summary reports of the most relevant and significant finds are given, presented by material type and chronology. Full reports can be found in White (2020), along with those on the post-prehistoric pottery, ceramic building material, metal objects, clay tobacco pipe and glass (mostly recovered from the soil overburden). Full catalogues of all materials can also be found in the archive (*ibid.*)

All artefacts were examined by appropriate specialists, who assessed the significance of the material in relation to the site and in terms of its wider importance. Quantified assessments were carried out for all artefact types and catalogues were produced.

Prehistoric Pottery

SARAH PERCIVAL

A total of 268 sherds weighing 2,354 g were collected from fifty contexts. The assemblage included sherds from a minimum of nine Early Neolithic Mildenhall-style bowls from a pit cluster in Area B of the site and Middle Bronze Age Deverel-Rimbury and Later Bronze Age and earlier Iron Age Post Deverel-Rimbury pottery from ditches associated with the enclosure (Table 2). The pottery was moderately well-preserved with an overall average sherd weight of 10 g (although this included larger sherds from the more robust Deverel-Rimbury urns).

Methodology

The assemblage was analysed in accordance with the guidelines for analysis and publication recommended by the Prehistoric Ceramic Research Group (2010). The total assemblage was studied and a full catalogue prepared. The sherds were examined using a binocular microscope (x10 magnification) and were divided into fabric groups defined on the basis of inclusion types. Vessel form was recorded and the sherds were counted and weighed to the nearest whole gram. Decoration, condition, food residues and sooting were also noted.

Early Neolithic

The Early Neolithic assemblage comprised 130 sherds (1,604 g) and included rims from nine vessels. All fabrics contained angular crushed flint exhibiting a range of abundance and inclusion size (Table 3). A range of vessels were represented, most with unshouldered profiles and heavy thickened rims (six of the nine examples). Typical was a bowl with a baggy profile and externally thickened rim

Table 2. Prehistoric pottery by spot date showing minimum number of vessels by rim count

Spot date	Quantity	Weight (g)	Minimum number of vessels	Mean sherd weight
Early Neolithic	130	1,604	9	12 g
Middle Bronze Age	43	698	4	16 g
Later Bronze Age	14	104	1	7 g
Earlier Iron Age	80	447	2	5 g
Not Closely Datable	1	1		1 g

(Figure 20, P1). Other forms included a small cup with a direct rounded rim, similar to vessels found at Kilverstone (Figure 20, P2) (Garrow et al. 2006: fig. 2.26, P42) and a third bowl with an everted rim (Figure 20, P3).

A single decorated vessel from pit (623) had incised lines forming a herringbone motif along the top of the rim which was externally thickened, both these traits being diagnostic of Mildenhall Ware (Figure 20, P4). The rim was similar to examples found locally at Spong Hill, North Elmham (Healy 1988: fig. 68, P86 and P89) and in abundance at Kilverstone near Thetford (Garrow et al. 2006: fig. 2.24, P51). Both Spong Hill and Kilverstone produced assemblages which contained decorated Mildenhall Ware vessels alongside contemporary undecorated examples in a range of forms, notably closed as well as neutral and sometimes open forms (Whittle et al. 2011: 763). One sherd had a post-firing piercing (Garrow et al. 2006: fig. 2.27, P35) perhaps undertaken as a repair on a vertical crack down the body of the bowl by binding through holes drilled on either side.

Middle Bronze Age

Middle Bronze Age Deverel-Rimbury pottery was recovered from ditch fills associated with the enclosure. The assemblage comprised a total of forty-three sherds weighing 698 g and included rims from four vessels (Table 4). The sherds were predominantly grog tempered with a small quantity being flint tempered. The temper is often coarse and blocky with inclusions visible breaking through the sherds surfaces. The sherds were frequently robust and chunky.

Rims were direct rounded (Figure 21, P6, P10) or direct flattened (Figure 21, P11) and derive from bucket and barrel-shaped vessels similar to examples found at Grimes Graves (Longworth et al. 1988: fig. 23) and Horsford (Percival 2018). Three sherds had pinched cordons (Figure 21, P7, P8), one decorated with fingertip impressions (Figure 21, P14), a decorative trait also present at Grimes Graves, Witton and Chimney Farm Witton (Longworth et al. 1988: fig. 25; Lawson 1983: fig. 25, 6; Gilmour et al. 2014). One rim was decorated with a row of fingernail impressions beneath the exterior of the rim (Figure 21, P11) (Lawson 1983: fig. 25, 9). Bases were fairly straight sided (Figure 21, P5) compatible with the tub and barrel-shaped vessels, with one example having a possible incised mark on the base (Figure 21, P9), perhaps made accidently whilst the urn was being formed.

Two radiocarbon dates were achieved on contexts which contained pottery. One, from enclosure ditch section (361) which contained a distinctive sherd with fingertip impressed cordon (Figure 21, P14) produced a date of 1497–1305 cal. BC (Beta 552729, 95.4 per cent). A second date of 1501–1323 cal. BC (Beta 552731; 95.4 per cent) came from enclosure ditch section (103) which also contained a flint and grog-tempered body sherd.

Table 3. Early Neolithic pottery by fabric

Fabric	Description	Quantity	Weight (g)	Number vessels by rim
F1	Sparse to common fine angular crushed burnt flint <0.25 mm in fine clay matrix	13	114	
F2	Sparse to common medium angular crushed burnt flint <0.25 mm–1mm in fine clay matrix	20	326	
F3	Sparse to common coarse angular crushed burnt flint 1 mm–3 mm in fine clay matrix	35	590	3
Q	Moderate fine rounded quartz sand	2	3	
QF1	Sparse to common fine angular crushed burnt flint in sandy clay matrix	26	376	2
QF2	Sparse to common medium angular crushed burnt flint in sandy clay matrix	18	116	3
QF3	Sparse to common coarse angular crushed burnt flint in sandy clay matrix	12	46	
QFvoids	Sparse to common fine angular crushed burnt flint with occasional sub-angular voids 2 mm–3 mm	4	33	1
Total	130	1,604	9	

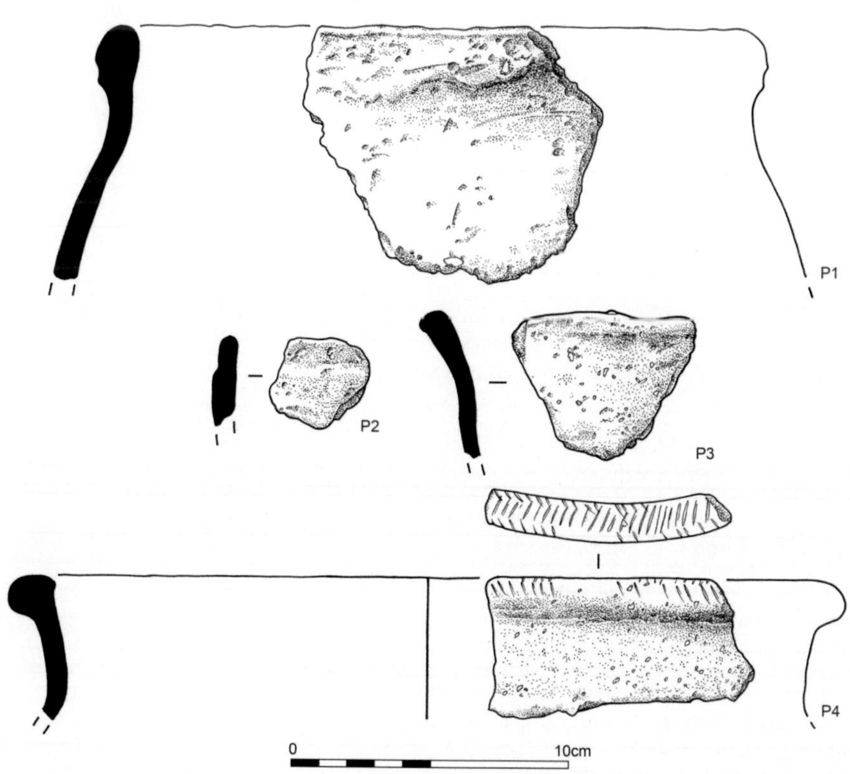

Figure 20. Early Neolithic pottery, selected sample. © c/o NPS Archaeology.

Late Bronze Age to Earlier Iron Age

A total of ninety-four Post Deverel-Rimbury sherds weighing 551 g were recovered from eighteen excavated features, mostly ditches forming the enclosure. A small quantity of sherds were unstratified. A number of sherds (fourteen sherds, 104 g) are typologically Late Bronze Age (1200–700 BC). The remainder had decorative and formal traits indicating an earlier Iron Age date (800–200 BC). The Post Deverel-Rimbury pottery was mainly made of flint-tempered fabrics which formed 94 per cent of the total assemblage by weight. Very small quantities of sandy and shell-tempered fabrics were also recovered (Table 5).

The Later Bronze Age sherds were characterised by the presence of distinctive thin-bodied vessel sherds with finger-wiped surfaces. A Late Bronze Age rim from enclosure ditch (776) was from a bipartite or barrel-shaped jar with in-turned neck (Brudenell 2012: form E2), similar to examples found on the Aylsham Bypass (*ibid.*: fig. 4.1 e2, d). Further Late Bronze Age sherds came from enclosure ditch (550).

The earlier Iron Age assemblage included rims from a round shouldered jar with an everted rim (Figure 22, P13) (*ibid.*: form A) and a small hemispherical cup (Figure 22, P12) (*ibid.*: fig. 4.1, form R). One base sherd had fingernail impressions on the interior. These are probably not decorative but were made whilst the vessel was being formed. One sherd from enclosure ditch (501) with deliberate decoration had multiple single fingernail impressions to the exterior. The sherd was similar to vessels found at Beeston Quarry (Ashwin and Flitcroft 1999: 244, fig. 23, P34) and West Harling (Clark and Fell 1953: 18, fig. 12, no. 26). Fingernail decorated jars of this type are dated to the Earliest to Early Iron Age and have a distribution centred on northern East Anglia (Brudenell 2012: fig. 6.25).

The distribution of the Post Deverel-Rimbury sherds is of interest. All of the Late Bronze Age pottery came from the enclosure ditch, whilst the Earlier Iron Age pottery came from Linear 1 (701=751) and from unstratified surface collection as well as several fills of the enclosure ditch. Four contexts within the enclosure produced mixed pottery assemblages; (287), (301) and (389). Each contained a mix of Middle Bronze Age and Earlier Iron Age pottery, whilst (776) contained both Later Bronze Age and Early Iron Age pot.

Table 4. Middle Bronze Age pottery by fabric

Fabric	Description	Quantity	Weight (g)	Number vessels
F4	Moderate coarse angular burnt flint 3 mm > in fine clay matrix	1	43	1
FG2	Moderate medium angular burnt flint >0.25 mm–1 mm, moderate sub-angular pale grog c.1 mm in fine clay matrix	6	128	
G1	Moderate to common fine sub-angular pale grog c.0.25 mm in fine clay matrix	12	29	
G2	Moderate to common medium sub-angular pale grog >0.25 mm–1 mm in fine clay matrix	16	303	
GC1	Moderate to common fine sub-angular pale grog c.0.25 mm in fine clay matrix with occasional medium sub-angular chalk and chalk voids	5	35	1
QFG3	Moderate medium angular burnt flint >0.25 mm–1 mm, moderate sub-angular pale grog c.1 mm in sandy clay matrix	2	155	1
QG	Moderate sub-angular pale grog c.1 mm in sandy clay matrix	1	5	1
Total		43	698	4

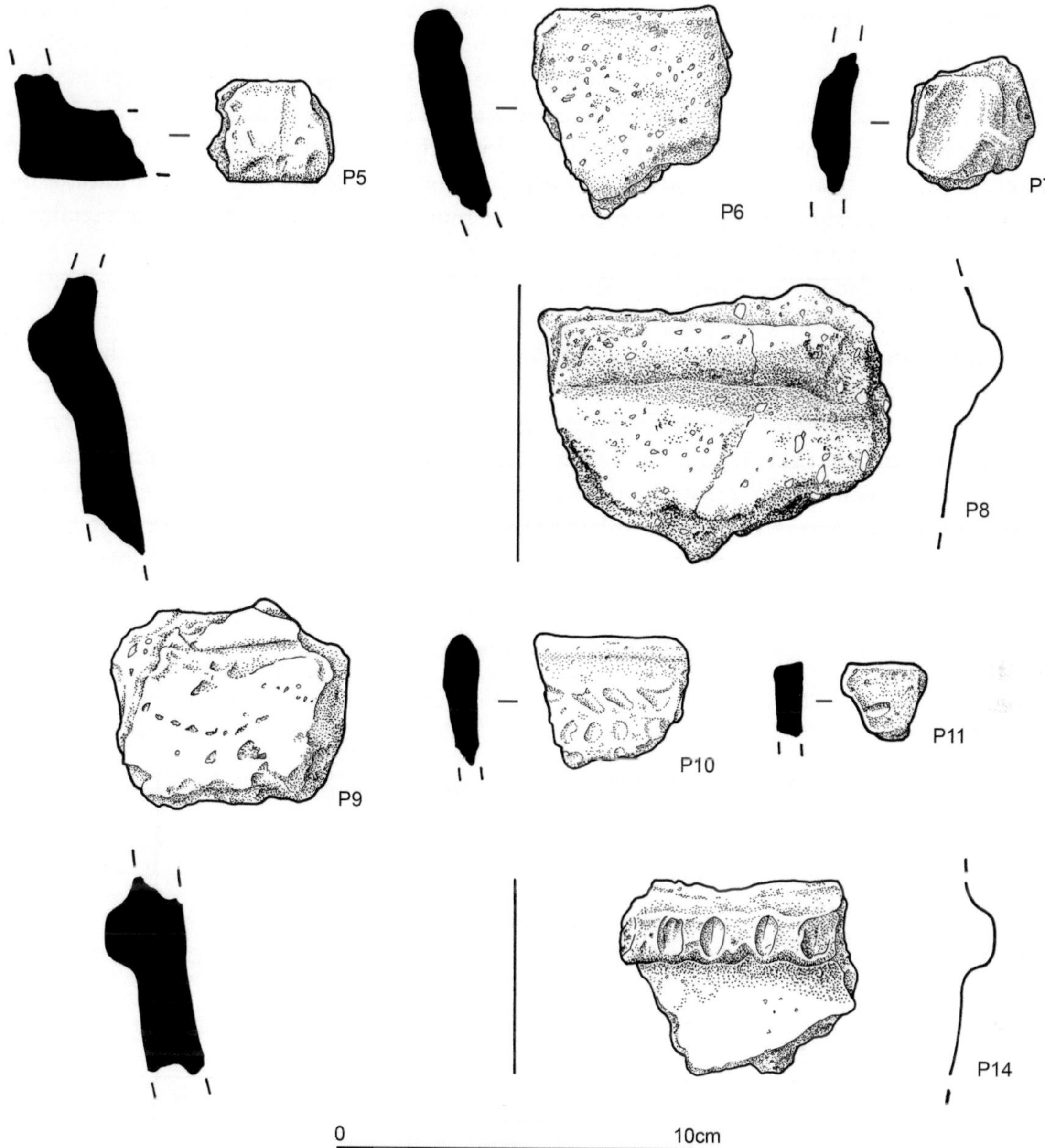

Figure 21. Middle Bronze Age pottery, selected sample. © c/o NPS Archaeology.

Discussion

The pottery assemblage indicates an episode of occupation activity in the Early Neolithic when Mildenhall type Decorated Bowl and associated Plain Bowl was deposited in a series of closely spaced and intercutting pits forming a cluster (Pit Cluster 1). The Early Neolithic assemblage comprised the fragmented and incomplete remains of at least nine vessels, dispersed through the fills of the pits. A radiocarbon determination on a nutshell recovered from pit (610) provided a date of 3645–3521 cal. BC (Beta – 552730, 95.4 per cent). Dating of Decorated Bowl pottery has been estimated to have a start date in south-east Britain of 3780–3685 cal. BC (95 per cent probability), probably 3745–3700 cal. BC (68 per cent probability) and an end date of 3310–3185 cal. BC (95 per cent probability), probably 3295–3225 cal. BC (68 per cent probability) (Whittle et al. 2011: 763). This places the Swan's Nest Neolithic activity towards the beginning of the proposed currency for Decorated Bowl.

The pit cluster from which Early Neolithic pottery was recovered at Swan's Nest recalls several of the 'developed' pit clusters found at Kilverstone, with the pits being irregularly shaped, tightly spaced and intercutting (Garrow et al. 2005; Garrow et al. 2006: 14). The pit clusters at Kilverstone produced a similar mix of Mildenhall type decorated bowl and contemporary undecorated bowl. A series of radiocarbon dates for Kilverstone were almost all very tightly grouped within a time bracket of 3650–3350 cal. BC (Garrow et al. 2005: 154). The radiocarbon date for Swan's Nest falls towards the beginning of the occupation period proposed at Kilverstone and produced a comparable assemblage probably indicating a similar occupation.

The Middle Bronze Age assemblage, and indeed the enclosure which produced it, are one of a growing body of Deverel-Rimbury activity identified in the county (Gilmour et al. 2014). The admixing of Middle Bronze Age Deverel-Rimbury and Post Deverel-Rimbury pottery within the ditch is intriguing and is similar to patterns noted at several sites in Norfolk, particularly at Horsford where features contained Middle and Later Bronze Age pot, though here the radiocarbon dates confirmed traditional dating of *c.*1422–1260 BC and 1742–1559 BC for the Middle Bonze Age pottery and of 898–802 cal. BC and 972–823 cal. BC for the Later Bronze Age (Tom Phillips pers. comm.). A mix of Middle and Later Bronze Age pottery similar to that identified at Swan's Nest and Horsford was also recovered locally in small quantities from an enclosure at Postwick Hub (Ames and Sillwood 2010;

Table 5. Post Deverel-Rimbury pottery by fabric

Spot date	Fabric	Description	Quantity		Weight (g)	Number of vessels
Late Bronze Age	QF1	Moderate fine angular crushed burnt flint (>0.25 mm) in sandy clay matrix	8		42	
	F13	Moderate to common coarse angular crushed burnt flint 1 mm–3 mm	6		62	1
Earlier Iron Age	F13	Moderate to common coarse angular crushed burnt flint 1 mm–3 mm	2		15	2
	QF1	Moderate fine angular crushed burnt flint (>0.25 mm) in sandy clay matrix	48		263	
	QF2	Moderate fine angular crushed burnt flint (>0.25 mm) in sandy clay matrix	13		71	
	F10	Moderate fine angular crushed burnt flint (>0.25 mm) in sandy clay matrix	9		59	
	Q2	Moderate round quartz sand	1		6	
	S1	Abundant very coarse shell	6		32	
Total			87		518	3

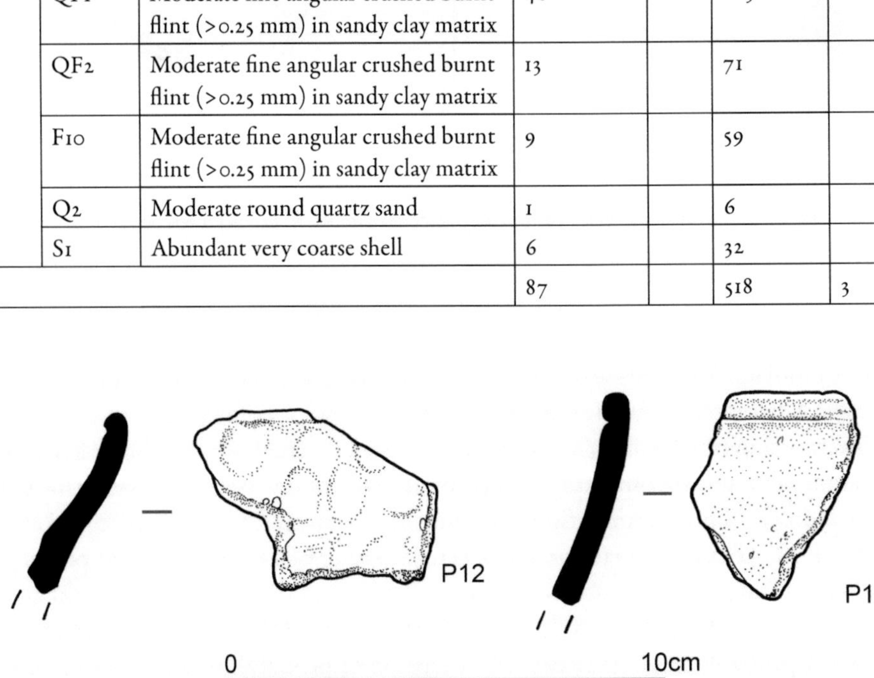

Figure 22. Later Bronze Age to earlier Iron Age pottery, selected sample. © c/o NPS Archaeology.

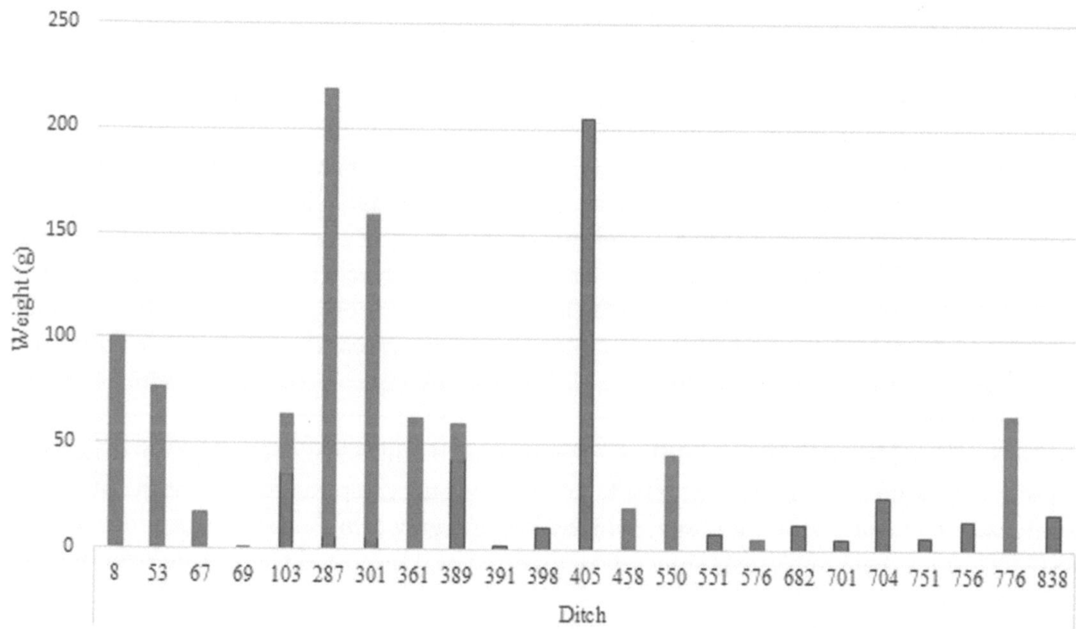

Figure 23. Middle Bronze Age and Post Deverel-Rimbury pottery from the enclosure ditches by intervention. Middle Bronze Age green; Later Bronze Age orange; and Early Iron Age blue. © c/o NPS Archaeology.

Crawley 2014). The admixing of Middle Bronze Age Deverel-Rimbury and Post Deverel-Rimbury pottery suggests that ditches forming the Swan's Nest enclosure were open for some time with a range of material becoming incorporated.

Trial trenching at Swan's Nest in 2016 produced a small assemblage of twenty-two sherds of pottery weighing 224 g. Fifteen sand and flint-tempered sherds were collected from the middle and 'lower' fills of the enclosure ditch including a handmade, barrel-shaped jar from the middle fills which could potentially also be of Mid to Later Bronze Age transitional date, though tentatively identified as being Iron Age (Peachey 2016a: 30). Also present in the evaluation assemblage were seven wheel-thrown or slow wheel formed Late Iron Age transitional forms (*ibid.*). No Late Iron Age to Early Romano-British pottery was subsequently recovered from the enclosure ditch during excavation. This is curious given the relative abundance of Romano-British finds in the vicinity and perhaps suggests that at least some areas of the enclosure remained open for some time after the earlier Iron Age and continued to receive material into the upper fills up to or after the Romano-British period.

The ceramic sequence for the material recovered from the enclosure, using traditional pottery dating, indicates that its use potentially began in around 1600 BC and persisted, whether intermittently or continuously, until at least the 800s BC. This traditional dating may, however, be under review. In Kent admixing of pottery with Deverel-Rimbury and Post Deverel-Rimbury traits has also been noted in features excavated along the line of HS2 and a hybrid mid/later Bronze Age ceramic phase proposed, characterised by ovoid jars in grog-tempered fabrics alongside Post Deverel-Rimbury forms in sandier fabrics (Champion 2011: 158). This hybrid phase is dated to *c.*1350–1000 BC (*ibid.*: table 4.1) a date which corresponds very well with the radiocarbon date achieved on juvenile cattle burial SK461 at Swan's Nest of 1386–1128 cal. BC (Beta – 552732, 95.4 per cent). It may be, therefore, that Swan's Nest provides evidence for the presence of a similar hybridised ceramic phase hinted at for the assemblage from the Norwich Northern Distributor Route at Horsford (Percival 2018) and possibly present at Ormesby St Michael (Gilmour et al. 2014).

Struck Flint

ANDREW PEACHEY

The excavation recovered a total of 188 pieces (3,059 g) of struck flint and a single hammer stone (871 g), including a range of retouched implements, blades and debitage (Table 6) that are largely characteristic of the technology of the Early Neolithic, with a tranchet adze and possibly several blades also indicative of Mesolithic activity. The potential Mesolithic pieces not only contrast slightly on technological grounds but were also heavily patinated, in comparison to the bulk of the assemblage which is largely un-patinated, notably a group of forty-seven pieces that were recovered from an intercutting cluster of pits dated to the Early Neolithic (Pit Cluster 1), that account for 25 per cent of the assemblage.

The implements and flakes recorded were manufactured utilising high-quality mid- to dark grey flint with a medium off-white to white chalky cortex. The only exception to this was the Mesolithic tranchet adze contained in Linear 1 (697), which while mid-grey, exhibited sparse to moderate round cherty inclusions (2–18 mm), suggesting it was imported or transported from an alternate and possibly non-local source.

Methodology

The flint was quantified by fragment count and weight (g). Flake type (see 'Dorsal cortex', below) or implement type, patination, colour and condition were also recorded. The term 'cortex' refers to the natural weathered exterior surface of a piece of flint, and the term 'patination' to the colouration of a flaked surface exposed by human or natural agency. Dorsal cortex is categorised after Andrefsky (2005: 104 and 115) with 'primary flake' referring to those with cortex covering 100 per cent of the dorsal face; 'secondary flake' with 50–99 per cent; 'tertiary' with 1–49 per cent and 'un-corticated' to those with no dorsal cortex. A 'blade' is defined as an elongated flake whose length is at least twice as great as its breadth, often exhibiting parallel dorsal flake scars (a feature that can assist in the identification of broken blades that, by definition, have an indeterminate length/breadth ratio). Terms used to describe implement and core types follow the system adopted by Healy (1988: 48–9).

Table 6. Quantification of the struck flint

Implement/Flake type	Frequency	Weight (g)
Tranchet Adze	1	162
Hammerstone	1	871
Backed Knife	1	75
End Scraper	6	273
Side Scraper	1	17
Side-End Scraper	2	50
Thumbnail Scraper	1	4
Blade	15	97
Platform Rejuvenation Flake	1	43
Debitage	160	2,338
Total	189	3,930

Commentary on Implements and Technology

Mesolithic

Linear 1 (697), fill (700), contained a complete, undamaged tranchet adze (Figure 26, F01); one of the core tools characteristic of the Mesolithic period in Britain. It was of a medium size and has been bi-facially-flaked resulting in a slightly convex upper face and a fairly flat underside. It had a lenticular section for much of its length but the tip had a slightly crested upper side, with a small tranchet flake removed from the underside to sharpen it. There is chipped wear to the cutting edge, but no evidence that it had been re-sharpened, or that damage had compromised its functionality.

Closely comparable tranchet adzes have been recorded at Little Dunham (Jacobi 1984: 71) *c*.5 km to the north-east, and at Great Melton (Wymer and Robins 1995: 139, fig. 11.100) *c*.20 km to the east. That there is a significant concentration of tranchet axes/adzes in west Norfolk has long been appreciated (Wymer 1977: 204–15; Jacobi 1984: 72, fig. 4.18), and this distribution includes further examples a short distance to the south-west at Gooderstone (NHER 20942; NHER 20943), to the north at South Acre and Castle Acre (NHER 4108; NHER 22083; NHER 3449). There is also a Mesolithic 'flint industry' site identified at Castle Acre (NHER 3890), where a high number of cores were associated with blades, microliths, tranchet axes and sharpening flakes. The addition of a further tranchet axe to this corpus of material, as well as the blades below highlights how advantageous the Breckland flint was to supporting a substantive (semi?) nomadic, probably seasonally mobile (episodic) population that was able to efficiently exploit the natural resources of what would have been heavily forested terraces in the Breckland landscape (Murphy 1984: 20).

The assemblage did not contain any cores, but was dominated by blade-based technology that was prevalent through the Mesolithic and early Neolithic, declining thereafter. Within the blades recorded there was an apparent contrast between those of possible Mesolithic origin (seven blades) and those of near-certain early Neolithic date (eight blades). The potential Mesolithic examples exhibited heavily patinated white surfaces, were slightly narrower (10–15 mm; Figure 24) with neater parallel dorsal scars along their length; and a more common (typical) incidence of platform abrasion/preparation.

Notably, none of the potential Mesolithic examples occurred within the pit cluster well-dated to the Early Neolithic; however three examples of these narrow blades were contained in pit (829) (Figure 26, F02–04), with further examples in enclosure ditch sections (129), (550) (Figure 26, F05), (588) (Figure 26, F06) and (705) (Figure 26, F07). The examples in ditches (129) and (705) exhibited wear on both lateral edges, while the blade from ditch (550) had possible wear on one lateral edge; however, all blades in the ditches were likely residual and may have been rolled. Further to the blades, similar technology and preservation was apparent on a platform rejuvenation flake also from enclosure ditch section (550), as well as un-corticated flakes of blade-like debitage from enclosure ditch sections (732) and (756). The platform rejuvenation flake was sub-rectangular with unidirectional blade removals across the dorsal face; likely representing the side of a core that may have been removed to refresh an existing platform, or may represent a mis-hit or unintended fracture.

This suite of blades and related flakes was closely comparable to the Mesolithic flint industry identified at Great Melton (Wymer and Robins 1995: fig. 5.54–66 and fig. 12.109–11) and consistent with the pattern of activity identified above in relation to the tranchet adze in west Norfolk (c.f. Wymer 1977: 204–15).

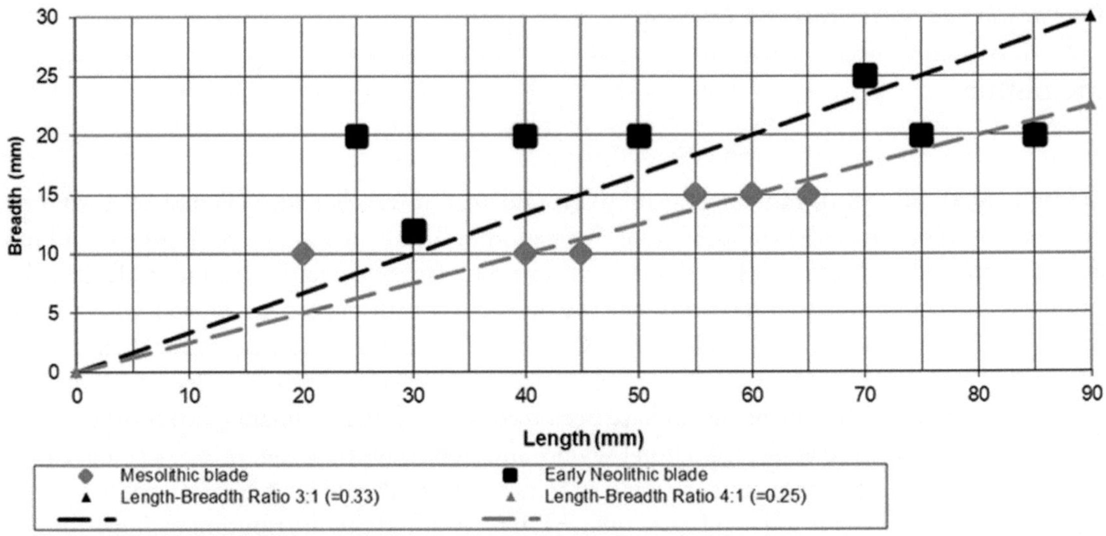

Figure 24. Comparative proportions of Mesolithic and Early Neolithic blades. © c/o NPS Archaeology.

Early Neolithic

The remaining blades in the assemblage were predominantly contained in the pit cluster dated to the Early Neolithic (Pit Cluster 1); in particular pits (597) (Figure 23, F08–09) and (627) (Figure 26, F10), with further examples as surface finds above them (621); and examples elsewhere from ditches (576), (652) and from the topsoil. Typically, these appeared slightly broader than the postulated Mesolithic examples, typically with a width of 20–25 mm (Figure 24). The bulk were also of a similar length, but the two examples from pit (597) were notable because they were longer (70 mm) with intriguing wear patterns.

One example had a limited extent of wear on a lateral edge close to the tip, which appeared both abraded and with edge gloss (Figure 26, F09), possibly resulting from cutting plant material. The second had fine chipped wear across the square distal end (Figure 26, F08), suggesting it may have been utilised similar to an end scraper or even a chisel; and it is notable that where present, the wear on Early Neolithic blades appeared more chipped than on the postulated Mesolithic examples and only ever on one edge. The production of longer, narrow blades, and the selection of flakes with straight edges with wear/polish limited to a restricted range of edge/tip angles was a feature of Early Neolithic groups at Spong Hill, where micro-analysis of edge gloss revealed a range of functions from cutting meat, wood, fibrous and green plant material (Healy 1988: 39 and 46).

The technology, size and proportions of the blades compare very closely to those in the large assemblage from Mousehold Heath, notably the groups accumulated in tree hollows and associated with blade production and slight variability in technology reflecting the specific temporal and functional needs of the knappers (Bishop 2011: 33 and 47); but these blades (and associated debitage, below) are only consistent with the narrower range of blades/flakes in pit groups at Kilverstone (Beadsmore 2006: 59, table 2.10), where a significantly broader range of core technology was associated with settlement, suggesting the activity that produced these blades represents a more expedient or episodic demand for blades.

Evidence for the modification of blades was very limited but included a single end scraper from pit (605) and a backed knife from enclosure ditch section (642). The backed knife (Figure 26, F11) was manufactured on a long blade, possibly utilising an early-stage removal from a large core and benefitting from one slightly blunt lateral edge. The opposing edge was heavily worn to the point that it is unclear if it was deliberately denticulated. Intensive pressure/abrasion had resulted in this edge being heavily chipped to the point that the natural grey flint has become opaque white on the leading edge only. Furthermore, the bulbar end had been narrowed by invasive retouch to form a small tang, suggesting the implement was hafted, potentially to function as a knife (or saw) with a carving or cutting motion. Although the probable tang remains unusual, the knife is otherwise comparable to types that occur in early Neolithic assemblages, including at Hurst Fen, Mildenhall (Clark et al. 1960: 222, F51 and F54) and Spong Hill (Healy 1988: 57, L86).

The remaining end scrapers, side scrapers and side-end scrapers were manufactured on elongate sub-ovoid flakes, often with blade-like dorsal scars, and with abrupt retouch generally only applied to the selected edges. These types of implements are common in assemblages that span the Early Neolithic in the region, including Hurst Fen, Mildenhall (Clark et al. 1960: 218) and Mousehold Heath (Bishop 2011: 40), with the end scrapers occurring entirely in the Early Neolithic pit cluster, including in pits (599) (Figure 26, F12) and (605) (Figure 26, F13–F15), as well as surface finds above them (contexts (620 and 622)); while this group also includes a side-end scraper in pit (607) (Figure 26, F16). The only slight deviation from this pattern in the pit group is a very large scraper (163g) in pit (605) (Figure 26, F15) which also utilises a sub-ovoid flake and has semi-invasive retouch around its curved distal end. It is unclear if this was manufactured on a deliberate flake blank, but perhaps may be regarded as a specialist tool, distinct from the typical range of end scrapers, but even such larger/thicker scrapers are within the range of tools at Hurst Fen (Clark et al. 1960: 218, F11).

In addition to the scraper types above, a single circular thumbnail scraper was contained in enclosure ditch section (592), in a slightly patinated and rolled condition that contrasts with the remainder of the assemblage. It was manufactured on a small primary flake, possibly an incidental chip. Thumbnail scrapers are most common in Early Bronze Age assemblages, and this may represent a casual loss or discard from that period.

The debitage flakes were dominated by tertiary and un-corticated flakes with blade-like proportions (Figure 25), and generally in the size range of 25–50 mm, not dissimilar to the blades. Sparse examples have abraded platforms and bulbs that appear to be the result of soft hammer percussion, while equally sparse examples are close to true blades, but the bulk are relatively undiagnostic beyond the fact they appear removed from systematic blade cores typical of the Early Neolithic. Relatively sparse flakes had elongate ovoid profiles, often with blade-like scars.

The debitage in the Early Neolithic pit cluster was predominantly of blade-like flakes, often associated with elongate sub-ovoid flakes (but not other flake types); and while consistently present across pits (597; 599; 601; 603; 605; 607; 609; 627; 629), the flakes were never present in any concentration (in total thirty-two flakes) that might suggest the presence of *in situ* knapping or the deliberate disposal/clearing of related waste/surfaces. Broadly the length-breadth ratios of these flakes were commensurate with the common types that account for over half the debitage in the pit groups associated with settlement at Kilverstone, with ratios between 0.2–0.4 and 0.4–0.6 (Beadsmore 2006: 58 and table 2.10), but this group is of insufficient scale to allow further meaningful statistical comparison, likely indicative of a brief expedient episode of knapping rather than the more consistent accumulation of waste from systematic core reduction.

Also from pits (627) and (629) was a large sub-rectangular secondary flake, which appeared to represent the hard-hammer percussion trimming of a core (or possibly rejuvenation); while a large blade-like tertiary flake in enclosure ditch (547) appeared to represent the similar removal of a crest or overhang on a core to facilitate the continued productivity of a striking platform.

The deliberate production of flakes, including broad-squat types may also have been a component of Early Neolithic core reduction; however, in this assemblage those present may represent core trimming and preparation, of possibly the later prehistoric use of flint. Small groups of hard-hammer struck, broad-squat debitage flakes were contained in enclosure ditch sections (588), (631) and (642) and may each represent the trimming, reduction or exploitation of a single nodule, although no cross-fits could be identified. The limited quantity of the groups suggests they do not represent significant episodes of knapping, perhaps rather the expedient trimming of nodules and cores, commensurate with that proposed for the Early Neolithic blades (above). This was likely part of a unified process, and if cores needed to be trimmed within this, then perhaps nodules were collected locally rather than transported as pre-prepared cores, supported by the ubiquity of available flint in the Breckland landscape.

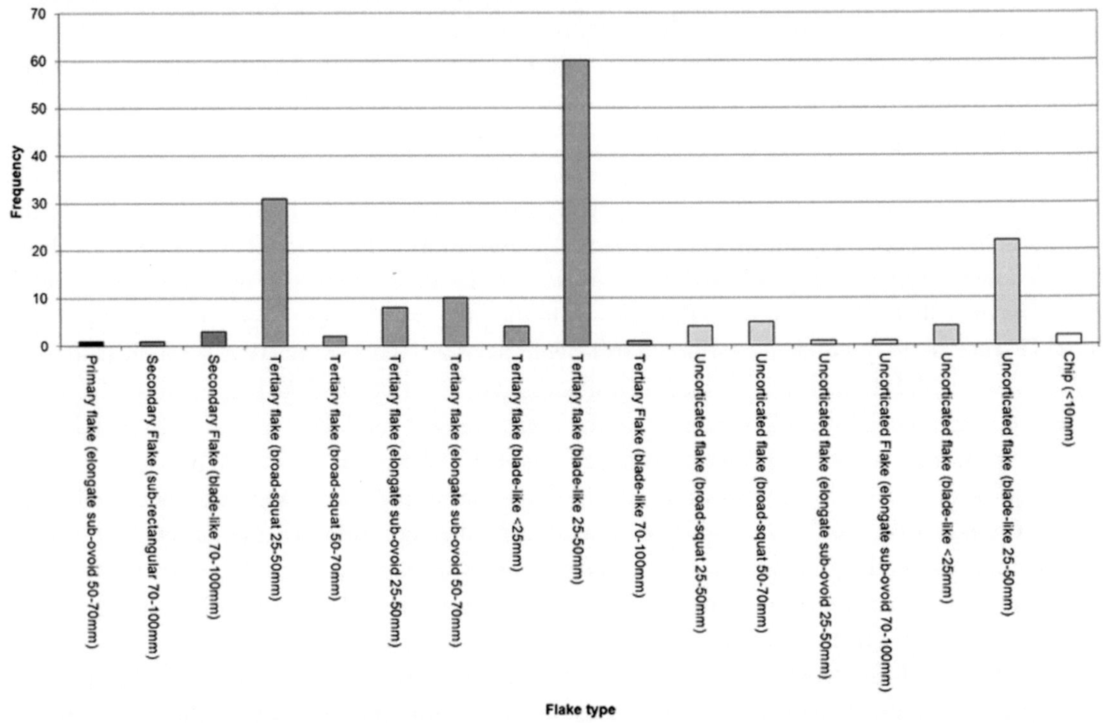

Figure 25. Frequency of debitage flakes by shape and size. © c/o NPS Archaeology.

The exploitation of systematic cores, characteristic of the Early Neolithic, to produce blades and associated flakes by the application of soft hammer, indirect percussion was further supported by a hammer stone contained in enclosure ditch section (301), fill (356). The hammer stone utilises a naturally spherical nodule of flint with all cortex intact (diameter *c*.85 mm), which is of comparable material to the knapped flint of Early Neolithic date. Based on the experimental comparisons of modern knappers in recreating the prehistoric toolkit, such a flint hammer stone (871 g) would classify as a 'large' (Whittaker 1994: 57) and 'hard' type (Lord 1993: 24). Yet this classification need not make the hammer stone a brutish, raw power tool; for its nature means it functions largely using its own weight with limited energy expended by the bearer, a technique that may be born out by the evidence of wear. One hemisphere of the hammer stone exhibits three lightly warn patches where the cortex has been lightly abraded and smoothed (but not worn through or chipped). This suggests the hammer stone was repeatedly and methodically used to strike an object softer than itself, such as an antler punch or fabricator used for the indirect, soft hammer percussion that allowed the accurate and repeated production of blades in the Early Neolithic (and Mesolithic).

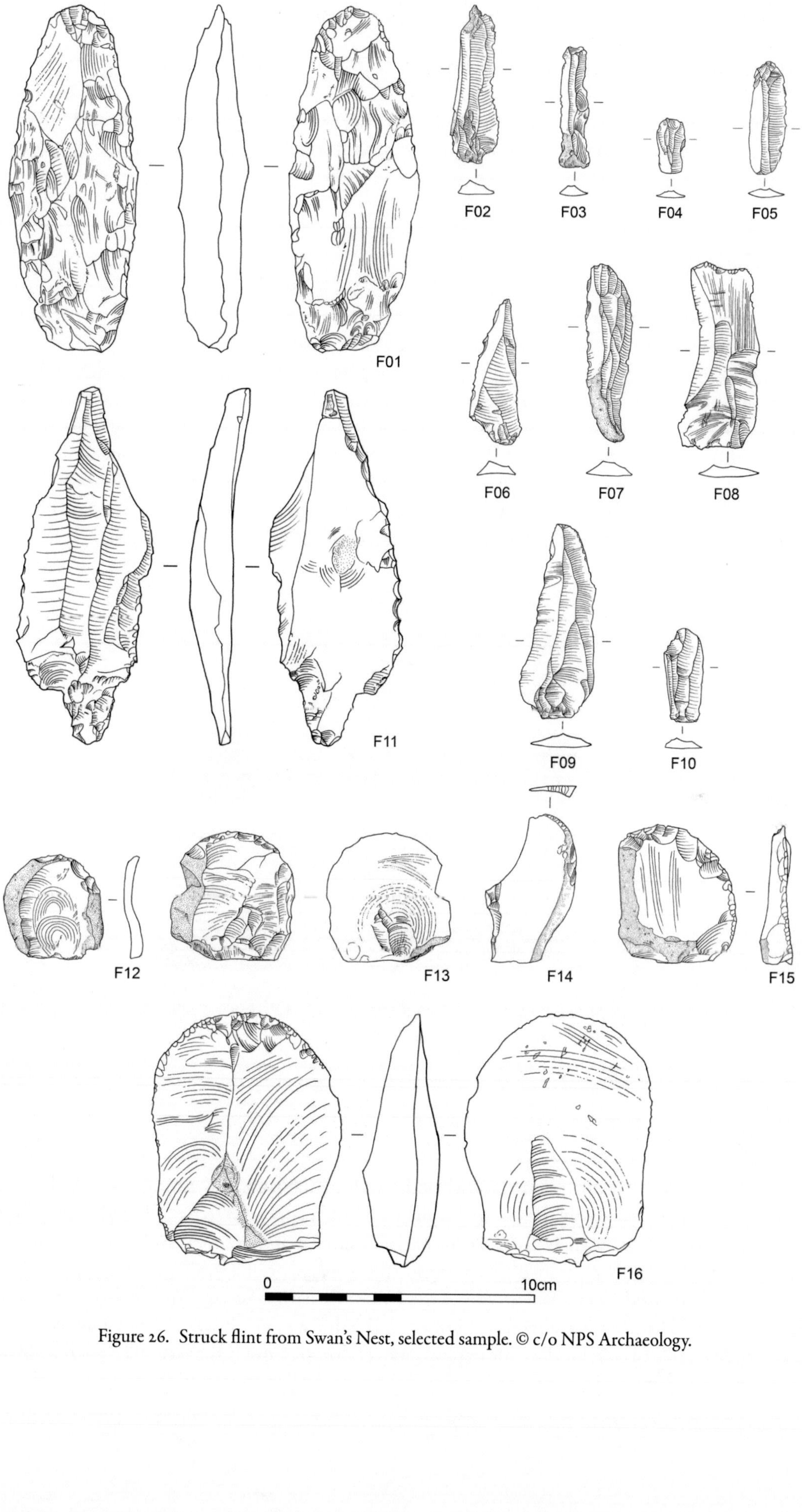

Figure 26. Struck flint from Swan's Nest, selected sample. © c/o NPS Archaeology.

Non-flint Stone

SIMON TIMBERLAKE AND FRANCES GREEN

A fragment of worked boulder along with thirteen other non-flint stones were collected during the excavation. Only the worked boulder and the five stones which exhibited the effects of heat, with sooting or reddened and cracked surfaces, are discussed here.

Worked Boulder

A small, apparently worked boulder of shelly limestone was found within fill (415) of entranceway ditch terminus (391) of the Middle Bronze Age enclosure (Plate 4).

A decision was made not to thin-section the stone on account of this being unlikely to be identifiable/provenanced using this method. Instead a full visual inspection of this specimen was made in hand, and the fossil clasts (shell fragments) compared to known fossil bivalves using standard palaeontological ID texts. A x10 illuminated table magnifying lens and a USB microscope were also used to help identify the fossils (by the former) and to confirm the presence of manual abrasion upon the grind surface (using the latter). A dropper bottle containing a solution of 10 per cent dilute HCL was used to confirm the presence of significant carbonate, confirming this as a limestone.

Plate 4. Grindstone/anvil from deposit (415), enclosure ditch terminus (391). Scale 10 cm. © c/o NPS Archaeology.

Identification

The rock was a massive shelly minor bioturbated micritic limestone containing flat-lying laminated layers of partially broken up bivalves – the majority of which appeared to be of Liostrea-type oysters. Only sections and a few fragments of the external shell surface of these were available to view, yet the likelihood is that these fossils were of *Praeexogyra* cf. *hebridica*, or of a similar *Liostrea* species; therefore most likely derived from a facies-type of the Upper Estuarine Series (Bathonian) of the Middle Jurassic (Natural History Museum 2001; Hudson and Palmer 1976). It was not possible to provide a certain confirmation of this, for the reasons given above, although this is the most likely determination.

The choice of this stone for use is almost certainly opportunistic, given that the most likely provenance for this stone in prehistory would be from amongst locally available glacial erratics brought down by the ice sheets and dumped within the local glacial till (boulder clay), and then redeposited perhaps within the river valley gravels. An original source for this rock therefore is likely to be north of the site along the coast/wolds of Eastern England, and possibly therefore North Lincolnshire or East Yorkshire.

Use as Worked Stone

The stone slab appeared to have been broken in use, or subsequent to its use, and is now probably just 50–60 per cent (?) of its original size. At the time of analysis, it weighed 4.86 kg and was more or less equidimensional across its top (200 mm × 200 mm) and approximately 80 mm thick.

The only surface which showed any evidence of working was the smooth flat top. It is important to note here that whilst this shows some evidence for having been used for grinding, the grind surface itself is poorly developed; that is, there are planated levels which are present across some of the tops of the original irregularities that show signs of having been worked in a specific direction, such as from the apex corner, yet the surface in all directions is not completely level, nor is it uniformly dished as one might expect with the longstanding use of such a stone as a saddlequern. Some of the working scratches formed by the action of a rubbing stone with this could however be seen on the surface detail. A further possibility is that apart from its relatively minor use as a grindstone, the centre of this selected slab may have been used as an anvil for the purposes of light crushing (though probably not for metalworking); a hypothesis which seems quite attractive in view of its incomplete development as a quern.

Thus, rather than referring to this as a saddlequern, with the consequent implications of this having been used for grain milling, it might be better to refer to this as a grindstone/anvil. There are in fact many examples of such dual-used stones from East England, particularly from the period of the earlier Iron Age, and it would seem likely therefore that this dates to this general period.

Carstone Lithology

One piece of carstone from deposit (07), the fill of Middle Bronze Age enclosure ditch section (08), showed signs of having been burnt. It was a poorly sorted orange, pale brown to dark grey sandstone, with a sub-rounded grit fraction and sub-angular coarse to fine sand. Opaque probably iron rich minerals were frequent and iron oxidised orange patches were common. The sandstone was clast supported and had variable degrees of cementation which makes this lithology friable as it weathers. Carstone is of Early Cretaceous date and occurs within the sequence of sandstones in the west and north of Norfolk. The closest outcrop is *c.*16 km to the west of Swaffham near Marham and was likely transported to the site in fluvial or glacio-fluvial outwash sands and gravels of Pleistocene date.

Greensand Lithology

There were two clasts of greensand; one from deposit (288), the fill of Middle Bronze Age enclosure ditch section (287) and one from deposit (112), the fill of Middle Bronze Age enclosure ditch section (103). Both of these stones had been used in a hearth or incidentally burnt in a fire. Under x40 magnification the sandstone was pale to white mainly well-sorted and fine grained with sub-rounded to sub-angular grains. The surface of the clasts was light grey with reddened patches. The stone was a well-cemented, well-sorted, fine-grained sandstone with sub-angular sand grains. It was clast supported with cement between grains hardly visible resulting in a hard rock with relatively high porosity. There were moderate, small, irregularly shaped dark brown to black inclusions (but not crystalline) which were probably organic or iron rich minerals. The greensand is of Lower Cretaceous date and outcrops in a band parallel to carstone some *c.*16 km to the west. It is likely to have been reworked in the Pleistocene by fluvial and glacio-fluvial processes.

Quartzitic Sandstone Lithology

One piece of quartzitic sandstone was recovered from deposit (187), the fill of Middle Bronze Age post-hole (186) (part of Fence-line 2). It was a pinkish brown angular fragment of a larger broken cobble. The outside of the stone was reddened and fractured as a result of heating. Away from the cortex the stone was brown and composed of well-sorted clast supported fine sand grains which were highly cemented with a dusty detrital cement with few opaque inclusions. Mica was rare and sparkled. It probably derives from the Triassic sandstones of the Midlands recycled in later river deposits draining from the west, such as the River Bytham.

Igneous Rocks; cf. Granite or Granodiorite, Diorite or Rhyolite

A burnt angular fragment of a sub-rounded cobble was found in fill (288) of Middle Bronze Age enclosure ditch section (287). The stone was dense and under x20 magnification no sand grains were obvious. It was a pale quartz in crystalline form together with larger dark brown lath shaped minerals. It was weathered and altered and was a possible igneous rock *cf.* granite or granodiorite but could be a metamorphic rock of some sort. It is likely to have been recycled through glacial activity and probably derives from outcrops in the north or the west of Britain, or even Scandinavia.

Heat-Altered Flint

JOSHUA C WHITE

A total of 587 pieces of heat-altered fire-crazed flint (25.6 kg) (hereafter burnt flint) were recovered from sixty-five separate contexts during the excavation. This material was recovered from Middle Bronze Age and earlier Iron Age features. The assemblage consisted of sub-angular pebble-sized fragments, their lengths ranging from between 5 mm and 110 mm. The fragments were largely of a mid- to dark greyish tone, with occasional specimens having a pale pinkish-red colour. Burnt flint is created through the heating of flint nodules in fires to high temperatures, and rapidly cooling it in liquid. This process results in the fracturing of the flint, while also changing its colour and giving it a fire-crazed appearance. Burnt flint has been recovered from secure Mesolithic (see Bishop 2018) to Anglo-Saxon (see Green 2008) dated deposits on other archaeological sites.

The most interesting burnt flint assemblages from the site (of which only a small representative sample was collected) came from deposits (380), (383), (385) and (388) of enclosure ditch section (361). These deposits appear to have been tipped in from the south of this feature, their matrix consisting of *c.*90 per cent burnt flint, with silt estimated to account for *c.*9 per cent and charcoal approximately

1 per cent. The small size of the burnt flint fragments that made up these deposits suggests that this material was heated on numerous occasions, fracturing it down repeatedly. Similar such deposits have been identified at Postwick, Norfolk (Crawley 2014) and Northwold, Norfolk (Crowson 2004), but rarely is such a density of flint recorded within the matrix of a deposit recorded.

Heat-altered flint was frequently recovered from other parts of the enclosure ditch beyond (361), although in much lower quantities. From the fills of the post-holes in Middle Bronze Age Fence-line 2, a total of 1.45 kg of burnt flint was recovered. A small quantity was recovered unstratified from above Early Neolithic Pit Cluster 1, although it is unlikely that it originates from the pits, as no heat-altered flint was directly recovered from any of their fills. See Chapter 6 for a full discussion of the burnt mound activity identified at Swan's Nest.

Fired Clay

SUE ANDERSON

There were fifty-one fragments (568 g) of fired clay recovered from thirteen contexts. Most of these were in fine sandy or silty fabrics, some with chalk inclusions and were generally of undiagnostic function. However, twenty-one fragments from deposit (789) of post-hole (788) (Post-hole Group 3) were in a medium sandy fabric with flint and voids, and the larger pieces were flattish with rounded corners, suggesting that these fragments were part of a triangular loomweight of Iron Age date. A similar such loomweight was recovered during the trial trenching (Peachey 2016b).

CHAPTER 4

Faunal and Environmental Evidence

Animal Bone

JULIE CURL

A total of 971 pieces (13,650 g) of animal bone were recovered during the excavation at Swan's Nest from Early Neolithic, Middle Bronze Age–earlier Iron Age and post-medieval dated contexts. Most of the faunal remains were collected from Middle Bronze Age–earlier Iron Age contexts, accounting for 94.7 per cent of the total assemblage (by weight). Due to the scarcity of datable finds from the deposits infilling the enclosure ditch, it has not been possible to consistently differentiate between the Middle Bronze Age, Late Bronze Age and earlier Iron Age animal bone, and thus the remains from these periods has been presented collectively. The post-medieval assemblage (accounting for 5 per cent by weight), which consisted of butchered cattle and equid remains, has been omitted from this dissemination and can be found in Curl (2020a).

Methodology

The assessment was carried out following a modified version of the guidelines set out by English Heritage (Davis 1992) and Baker and Worley (2014). All of the bone was examined to determine range of species and elements present. A record was also made of butchering and any indications of skinning, hornworking and other modifications.

Where possible, ages were estimated along with any other relevant information, such as pathologies. Counts and weights were noted for each context and counts made for each species. Where bone could not be identified to species, they were grouped as, for example, 'large mammal', 'bird' or 'small mammal'. Minimal measurable bone was recovered, with two limb bones recorded following Von den Driesch (1976). A full catalogue of the faunal remains is available in the archive (White 2020).

The Assemblage

Quantification, Provenance and Preservation

An assemblage of 12,954 g of bone, consisting of 939 pieces, was hand-collected from prehistoric deposits at the site and is quantified by feature, weights and counts in Table 7. The sieved samples produced a further 12 g of bone consisting of nineteen fragments (Table 9). Remains were mostly produced from the fills of the Middle Bronze Age–earlier Iron Age enclosure ditches, which amounted to 97 per cent of the assemblage by weight. A small quantity of bone was recovered from pit fills at the site. A small assemblage of bone was collected from deposits dated to the Early Neolithic (0.3 per cent of the total assemblage by weight).

The bone was generally in good condition, although was quite heavily fragmented from butchering and wear. Despite the age of the remains, butchery marks were seen across much of the assemblage. One complete skeleton was recovered which was in a very good condition, with numerous small elements present. A few fragments of bone showed greater wear and erosion than other bones from the same context, suggesting a few pieces had been residually deposited, but their numbers are low and the vast majority of bone is considered to have been recovered from its original place of deposition. Invertebrate damage was seen on bone from a few of the enclosure ditch fills, which would suggest that some remains were left partly buried or uncovered, allowing insects, isopods and molluscs to feed on the bones and waste. Canid gnawing was observed on bones from two deposits. Cattle metatarsals from the fills of enclosure ditch sections (405) and (551) both show gnawing at the ends of the bone, which might suggest skinning waste was available for scavengers.

Table 7. Quantification of the hand-collected faunal remains by feature type, weights and counts

Feature type	Ditch	Ditch (458) with SK	Pit	U/S from (405)	Grand total
Total weights	9,549	3,195	75	135	12,954
Total counts	684	229	11	15	939

Table 8. Quantification of the faunal remains by feature type, species and NISP

Species	Feature type and NISP				Totals
	Ditch	Ditch (458) with SK461	Pit	U/S from (405)	
Cattle	286	229	11	15	541
Deer – Roe	1				1
Deer – Red	1				1
Deer/sheep/goat	1				1
Equid	7				7
Mammal	356				356
Pig/boar	6				6
Sheep/goat	25				25
Vole – Bank	1				1
Totals	684	229	11	15	939

Species Range

Eight species were identified in the hand-collected assemblage, which are quantified in Table 8. The assemblage produced mostly domestic food stock with cattle, pig, sheep and goat, along with small amounts of equid. Wild species were represented by deer and a vole.

Pathologies

No pathologies were identified, and it seems the stock at this site were healthy. None of the animals in this assemblage showed any great age, which would also reduce incidences of health problems.

Early Neolithic

Fragmented juvenile cattle teeth (35 g) were recovered from pit (599) and a fragment of cattle tibia was recovered from pit (601). Both pits were part of Pit Cluster 1. Little can be said about these poorly preserved fragments, but they do attest to the presence of cattle at the site during this period.

Middle Bronze Age–Earlier Iron Age

Cattle (*Bos taurus*) were the most frequent animal remains encountered, mostly coming from the fills of the enclosure ditches. Ages of the cattle varied, most were adult at death, others were sub-adult, juvenile and neonatal. One articulated animal was seen, while other deposits produced mostly limbs, with some scapula, pelvic fragments, vertebrae, ribs, head/horn and teeth. Butchering was frequently seen on many of the bovid remains.

The largest collection of cattle remains were produced from enclosure ditch terminus (458), with an articulated young animal SK461 in fill (462) (Plate 5). Fusion stages for the limb and foot bones were noted following Cornwall (1974) and they suggest an animal that died around 1.5 years old, so young, but not a sickly/weak calf dying shorty after being born. This animal was found with Middle Bronze Age pottery, and in remarkably good condition with many small elements (including intermediate and distal phalanges) surviving very well, so it is unlikely that even the finest of butchering evidence, if present, would be lost. A sample of bone was sent for radiocarbon dating from this individual, which produced a date of 1386–1128 cal. BC (Beta – 552732, 95.4 per cent). The animal was laying on its left side with the forelegs folded back and the hind legs stretched out underneath it. A few butchering marks were noted, with small cuts on a proximal humerus, and the femurs had knife cuts on the proximal shafts. There were no typical cuts from skinning present.

Deposit (295) of enclosure ditch section (550) contained the remains of an adult cow and neonatal calf. The overall number of cattle bones seen in the analysis totalled 530 fragments. Removing SK461 from the overall cattle number, leaves 275 bovid elements, still making them the most frequent species at the site.

Little measurable bone was present in this assemblage. One cattle metatarsal from a Middle Bronze Age ditch fill suggested a shoulder height of 1,117 mm, which is in the range for breeds such as the Celtic Short-Horn, so indicating quite small animals and not a large beast more closely related to the Aurochs and its descendants, which did survive into the Bronze Age in Britain. The ages of cattle in this assemblage showed most variation, with adults, juveniles and neonatals present, indicating on-site breeding of cattle.

Sheep/goat remains were recovered from seventeen ditch fills, mostly with single bones or a small number of elements. Most remains were from quite small individuals and are likely to be from sheep (*Ovis aries*); however, a couple of bones did suggest the more robust goat (*Capra aegagrus hircus*). Analysis suggests that all the ovicaprid remains were from adults. Most remains were limbs, with some head elements. Deposit (376) of enclosure ditch section (361) produced one sheep horncore which had been chopped and cut, suggesting hornworking waste. Butchering was seen, but less frequently than with the cattle, possibly suggesting these animals were cooked whole and required less butchering than larger animals.

Pig/boar (*Sus scrofa domesticus*/*Sus scrofa*) remains were recovered from four ditch fills with limb, mandible and teeth elements present. All of the porcine remains were from juveniles, with some limb bones showing butchering. Five deposits produced equid remains, most were limbs, with upper molars seen in the enclosure ditch fill (331) of excavated section (301). Some butchering was seen, suggesting equids were at least skinned.

Cervids were represented by a Roe Deer (*Capreolus capreolus*) mandible in deposit (460) of enclosure ditch (458). A possible Roe Deer tibia was seen from deposit (413) of enclosure ditch (391),

Plate 5. SK461 in enclosure ditch terminus (458), facing south-east. Scale 0.5 m and 0.3 m. © c/o NPS Archaeology.

which had been butchered. A Red Deer (*Cervus elaphus*) tibia, which had been chopped, was produced from deposit (295) of enclosure ditch (550). A single Bank Vole (*Myodes glareolus*) pelvis was found in deposit (460) of enclosure ditch section (458), which shows excellent preservation. This species of rodent is commonly found living in and around ditches and may represent a resident of the ditch. These animals burrow and, given the preservation, it may be a modern animal, but a contemporary animal is quite possible given the good preservation of other bones at this site. Many species also eat voles, including mustelids, cats and birds of prey and remains may arrive in the ditch via kills from these species, particularly regurgitated bird pellets.

Sieved Samples

A further 12 g of bone, consisting of nineteen elements, was recovered from the flots of sieved samples. The bone is in good condition, although much of it is heavily fragmented. Bone was recovered from four ditch fills, all of Middle Bronze Age–earlier Iron Age date.

From Middle Bronze Age post-hole (439), sample <31>, two fragments of avian egg shell were recovered. A single fragment was also yielded from deposit (702) of earlier Iron Age Linear 1 (sample <57>). The avian egg shell is quite robust and at least one fragment resembles duck egg shell.

From deposit (62) of enclosure ditch section (53), sample <01> produced two metapodials from a Weasel (*Mustela nivalis*), the smallest species of the order Carnivora in Britain. A Weasel scapula was also recovered from deposit (388) of enclosure ditch section (361). This little mustelid, weighing less than 200 g for a male (48–107 g for a female), is a specialised predator of voles (Couzens et al.

2017) and hunts in the burrows of its prey. The Weasel's habitat is open ground, woodland and hedges, often hunting along the length of a hedge. While voles are a large part of their diet, this mustelid will take mice, small rabbits, birds and eggs. It has a high need for regular food, consuming a third of its body weight each day, and surplus prey is killed and cached for later consumption. Other bone from samples was only identifiable as 'mammal'.

Table 9. Quantification of the sieved sample material

Deposit	Sample no.	Type	Period	Context qty.	Wt (g)	Species	NISP
(62)	1	Ditch (53)	MBA–eIA	6	4	Weasel	2
(62)	1	Ditch (53)	MBA–eIA			Mammal	4
(388)	19	Ditch (361)	MBA–eIA	1	1	Weasel	1
(460)	29	Ditch (458)	MBA–eIA	8	6	Mammal	8
(545)	22	Ditch (550)	MBA–eIA	4	1	Mammal	4

Discussion

The bulk of the Middle Bronze Age–earlier Iron Age assemblage appears to be derived from the primary and secondary butchering and meat waste from the main stock animals, with a dominance of cattle. Some hunting is indicated by Roe Deer and Red Deer, with Wild Boar possibly represented in the porcine remains. The lack of bird is not surprising as few domestic birds were kept in prehistoric times and poor preservation can often affect survival of bird bone. Similarly, fragile fish, if consumed, might not have survived. It is possible, given the preservation of some small bones in this assemblage that these items were consumed and disposed of elsewhere, or not consumed at all.

The sieved samples provided an additional species with the Weasel, which is a hedgerow and woodland to rough grassland species and a predator of voles, with one vole found in the hand-collected faunal material also. Weasels are occasionally identified in archaeological deposits, with one found in a Romano-British deposit from Poundbury (Grimm 2011).

The Articulated Juvenile Cattle

The Middle Bronze Age cattle skeleton in the enclosure ditch terminus raises the question as to whether it was a burial of a 'ritual' nature or is simply the mundane disposal of a carcass. Juvenile cattle can be of a clumsy nature and are known to fall into ditches, where they can struggle to free themselves and have the potential to die. However, this is not believed to be the case with this young animal, which was notably situated high up in the stratigraphic sequence and thus was buried at a point when the enclosure ditches were only partially open (and not particularly deep). From the positioning of its limbs, it also appeared that it had been placed/dumped into the ditch, rather than having died while struggling to free itself. Most significantly however cut marks were identified on the animal, indicating human interaction, interference and responsibility for its deposition.

Cut marks may suggest that the animal was possibly skinned, but there is a distinct lack of butchering to the carcass suggesting that the animal was at least not extensively consumed. It may be possible to speculate that the young animal had died a natural death and was perhaps scavenged or partially decayed when found. Prior to being dumped into the partially filled enclosure ditch terminus, skinning and/or some investigation of the animal was carried out to see if any meat was suitable for consumption.

Alternatively, it may be possible that the young bovid was part of a more structured ritual and was sacrificed, with some prime meat then removed, prior to the animal being placed in the ditch. Complete animal skeletons of later prehistoric date are often interpreted as 'ritual' deposits, although more frequently when identified at the base of pits (see Smith 1965; Grant 1984; Grant 1989; Grimm 2011). A similar cattle specimen was found deposited in a pit on the threshold of the Middle Bronze Age enclosure at Aller in Somerset, which possibly represents a ritual deposit (Allen 2017). Hill (1996) discusses how interpretations of 'special animal deposits' have been controversial. In contexts where the repeated consistent placement of these deposits is observed on a site, and their positioning is such that it is both involved and deliberate, their association with ritual practices is on the whole, widely accepted. Cunliffe (2003: 149), for example, has speculated that these deposits may represent 'proprietary offerings' to 'chthonic deities'.

However, the specimen from Swan's Nest is unique for the site. Its positioning is also clearly not of a considered or respectful nature, with the animal having a distorted rear that appears to have snagged on the ditch slope, pushing its back legs forward, implying the carcass was either dumped or dragged with force. Consequently, the former scenario discussed above is favoured for the burial context of this juvenile animal.

Conclusions

The bulk of the assemblage from the Middle Bronze Age–earlier Iron Age activity at the site was derived from primary and secondary butchering and food waste from the main domestic stock animals (cattle, sheep, goat, pig and horse). Some supplementing of the diet from local wild stock occurred, including Red Deer and Roe Deer, which would have been available in nearby woodland. The assemblage includes a young bovid burial, which likely represents an animal which died naturally at pasture and was deposited into the partially filled enclosure ditch.

The assemblage is too small for meaningful analysis on kill patterns and full range of uses, but the neonatal cattle indicate on-site breeding, with perhaps a cull of some neonatals to allow the milking of the mothers. The young age of most porcine bone might suggest domestic bred pigs, but equally a juvenile wild boar is easier to hunt. The sheep/goat were all adults and this might suggest keeping for fleeces or milk was a greater need than prime meat. Environmental evidence was provided by the deer, which suggest nearby woodland and grassland for grazing, as do species like the Weasel and Bank Vole.

Molluscs

JULIE CURL

A total of 197 mollusc shells (NISP) were recovered from Middle Bronze Age–earlier Iron Age and post-medieval deposits during the excavation at Swan's Nest. The post-medieval remains have been omitted from the below dissemination and are detailed in the archive (Curl 2020b).

Methodology

The molluscs were identified to species using a variety of reference materials. Shells were catalogued by species and counts were made of the number of individual species present (NISP). Where appropriate, shells were examined for any cut marks that would confirm their use for food from the prising

Table 10. Quantification of the mollusc assemblage

Species	Ditch	Linear	Pit	Post-hole	Totals for species
Ceciliodes acicula				2	2
Clausilia bidentata	5			3	8
Clausilia sp.	4				4
Cochicopia sp.	1				1
Coclodina laminata	1				1
Discus rotudus	5		1	7	13
Ena obscura	10			8	18
Helicigona lapicida				1	1
Helix sp.				1	1
Lymnaea palustris	1				1
Misc. shell	56			4	60
Oxychilus alliarius	16			11	27
Ostrea edulis			3		3
Pupilla muscorum	5			1	6
Trochulus striolatus	11				11
Vallonia costata	1		1	1	3
Vallonia pulchella	6	1	2	2	11
Vallonia sp.	7		1	3	11
Vertigo pygmaea	3	1	1		5
Vertigo sp.			1		1
Cepaea hortensis	2				2
Zonitoides sp.	3				3
Totals by feature type	137	2	10	44	193

apart of shells or removal of meat with a knife. The shell assemblage from this site includes many small land molluscs retrieved from sieved samples.

Quantification, Provenance and Preservation

A total of 193 mollusc shells, amounting to at least 34 g (with additional weight from each context recording <1 g) were recovered. Shell was produced from a variety of ditch, post-hole and pit fills, and was associated mostly with ceramics of Middle Bronze Age to earlier Iron Age date. The shell is quantified by species and NISP in Table 10, with full quantification by individual contexts available in the archive (White 2020).

The land mollusc remains were in a very good condition; a factor possibly aided by the calcareous soils of the area. Some fragmentation had occurred, which is expected with processing and particularly with the marine molluscs. A total of seventeen individual land or wetland species/species groups were identified, with some fragments recorded as, for example, '*Vallonia* species'. A single marine species was identified.

Plate 6. A selection of the land molluscs found in Middle Bronze Age enclosure ditch (103), deposit (118), sample <02>. Photographed using a digital microscope. Scale 1 mm sections. © c/o NPS Archaeology.

Marine Molluscs

Three fragments of Common Oyster (*Ostrea edulis*) shell were recovered from sample <20> of Middle Bronze Age pit (211). They were small fragments and may potentially be intrusive.

Land Molluscs

Land molluscs were recovered from twenty-one Middle Bronze Age–earlier Iron Age sampled deposits, with sixteen species identified. These are discussed below by environment.

Burrowing/Intrusive

Ceciliodes acicula was seen from sample <23> of Middle Bronze Age post-hole (218) and from sample <33> of Middle Bronze Age post-hole (427). This is a subterranean species, a burrowing snail that can probably be classed as intrusive. It is possible that this species is a fairly recent introduction as there are no known records of this species in the fossil mollusc records of Britain (Evans 1972).

Widespread Species: Dry Calcareous Grassland

Pupilla muscorum were produced from samples of three enclosure ditch interventions and a post-hole of Gate Structure 1 and *Vallonia costata* were found in samples from an enclosure ditch intervention, a post-hole of Gate Structure 1 and a Middle Bronze Age pit; both are common in this environment and prefer shorter grassland. *Discus rotundus* is a widespread species found in a wide range of environments, which can include dry and rougher wet areas. This species was recovered from a number of enclosure ditch fills. A single fragment of a *Helix* sp. was retrieved from Middle Bronze Age post-hole (427) (Gate Structure 1), sample <21>, probably from *Helix aspersa*.

Grassland, Ditches, Wetter Environments

These species generally prefer rougher grasses and areas close to and inside ditches, but generally not permanently flooded ditches. *Vertigo pygmaea*, a grassland and wetland species, were found in sample <02> from Middle Bronze Age enclosure ditch (103) (Plate 6), and sample <22> from enclosure ditch (510), as well as from Middle Bronze Age pit (212) and linear (701). *Cochicopa* sp. was also seen in sample <02> and might be found around ditch margins (Plate 6).

Vallonia species were seen in samples from across the enclosure ditches. *Vallonia pulchella* was seen from pit (212), post-holes (456) and (793) and enclosure ditches (456), (756) and (776). These snails are primarily found in wetland areas and near ditches. A higher number of *Vallonia pulchella* was seen in sample <61> of Middle Bronze Age enclosure ditch (756), which would suggest the main source of this species was the enclosure ditches, with this species perhaps entering other features via heavy rain through flooding or via collected water.

The larger land molluscs consist of two shells of *Cepaeca hortensis*, also known as Garden Snail or White-Lipped Snail, which were discovered in sample <22> from enclosure ditch (510). This species is found in woodland, hedges and shrubs. Generally, it prefers wetter and cooler places than the related Brown-lipped Snail, so its presence in a ditch environment is expected. At just under 20 mm, these shells are close to the maximum adult size, so they probably died naturally at the end of their lives.

Woodland and Hedges

These species tend to favour woodland, hedges and wood in isolation, the latter including log piles or fallen trees. *Clausilia bidenta* were seen from numerous enclosure ditch samples as well as from post-holes of Gate Structure 1. This species can be found in these environments and under logs. Similarly, *Coclodina laminata* is also found under logs and in litter and was found in enclosure ditch (510). *Ena obscura* was seen in samples from enclosure ditches (456) and (510) as well as in a sample of Fence-line 2 post-hole (186), and particularly favours beech wood. *Trochilus striolatus* and *Oxychilus* sp. were recovered from numerous enclosure ditch fills, as well as from post-holes (428) and (447) of Gate Structure 1. Both are species of woods and hedges, but also of waste ground and more open areas.

Three shells from *Zonitoides* sp. were found from samples taken of enclosure ditch (510), fill (545). This is a snail of damp, woodland and marshy wood; perhaps suggesting a slightly wetter ditch area or wood that had fallen into the ditch.

Limestone, Hedge and Woodland

One shell from *Helicigonia lapicida* was produced from Gate Structure 1 post-hole (445), sample <32>. This species can be found in woodland or well-established hedges and on calcareous soils, but also favours limestone rock. In wooded areas, this species prefers older woodland and well-established hedges, perhaps suggesting a ditch that has been used for a long time with a well-established hedge, including developed trees close to it. This species was also seen in the trial trench evaluation (Summers 2016).

Discussion

The land molluscs from this site have provided good environmental information, with many species identified having quite specific habitat requirements. The majority of the shells came from ditch fills and consisted of species which occupy areas close to ditches, where rough grasses and shrubs are usually present. Many species identified can also be found around wood and hedges. Overall, evidence suggests that the enclosure ditches had long grass edges with shrubs and trees close by. Trees and shrubs along with well-established hedgerows would produce build-ups of leaf litter that many of the identified species would inhabit. The presence of some snails which prefer more rotting wood and logs may suggest supplies of prepared and stored wood. The log-loving species might suggest fallen trees, but it is likely that any fallen trees would be utilised for timber or firewood and would not have been left around the site to rot.

Plant Macrofossils

VAL FRYER

Forty-six bulk samples were taken for assessment of environmental evidence during the excavation. Of these, the following fourteen samples were selected for analysis based on their potential archaeological significance:

- Samples <47>, <48>, <49> and <50> from Early Neolithic Pit Cluster 1;
- Samples <16>, <17>, <18> and <19> from the Middle Bronze Age burnt flint deposit from enclosure ditch section (361);
- Sample <58>, a charcoal-rich deposit of Middle Bronze Age–earlier Iron Age date from enclosure ditch terminus (671);
- Samples <29> and <59> from Middle Bronze Age–earlier Iron Age dated deposits from the enclosure ditch sections (458) and (704) respectively;
- Sample <33> from Middle Bronze Age post-hole (428), which forms part of Gate Structure 1;
- Sample <55> from the Middle Bronze Age enclosure ditch section (642); and
- Sample <26> from Middle Bronze Age post-hole (176), which forms part of Fence-line 2.

It was proposed that analysis of the recovered assemblages would provide:

- Information about how the large enclosures and ancillary features may have been used;
- Data which could be compared with that from similar Middle Bronze Age 'Ormesby-type' enclosures; and
- Information about the local environment during the lifetime of the site.

Methodology

The samples were processed by manual water flotation/washover and the flots were collected in a 300 micron mesh sieve. The dried flots were sorted under a binocular microscope at magnifications up to x40 and the plant macrofossils and other remains noted are listed in the archive (White 2020). Identifications were made by comparison with modern reference specimens. All plant remains identified were charred.

Sample Composition

Cereal grains, seeds and nutshell fragments were present at a low to moderate density within nine of the assemblages studied during the analysis. Preservation was mostly poor; the cereal grains in particular were especially puffed, fragmented and abraded, possibly as a result of high temperature combustion and subsequent mechanical damage within the archaeological soil horizon. In addition, some macrofossils were heavily encrusted with fine grits and mineral concretions. Whilst the latter has not impeded identification, it is likely that it may have precluded the full retrieval of plant macrofossils during processing. However, the non-floating residues from the samples were carefully sorted and any obvious plant materials were returned to the appropriate flots.

A single elongated 'drop' form wheat (*Triticum* sp.) grain of possible spelt/emmer (*T. spelta/dicoccum*) type was present within the assemblage from post-hole (176), part of curved Fence-line 2. Other cereal grains were very scarce, and all were too poorly preserved for close identification. However, a possible emmer glume base, with a single prominent keel, was noted within the fill (388) of enclosure ditch section (361). Sample <49> from pit (609) of Early Neolithic Pit Cluster 1, included what appears to have been a cotyledon fragment of an indeterminate large pulse (*Fabaceae*), but the specimen was severely abraded, and further identification was not possible.

Weed seeds only occurred within the assemblage from enclosure ditch section (642). Both were extremely poorly preserved, but a fragment of goosegrass (*Galium aparine*) was noted along with a single indeterminate large grass (*Poaceae*) fruit. Hazel (*Corylus avellana*) nutshell fragments were present/common within three of the four Early Neolithic pit fills, and two further possible fragments were also noted within the Middle Bronze Age enclosure ditch section (361), deposit (376). Other plant macrofossils were very scarce, but an indeterminate bud, culm node and fruit stone/nutshell fragment were noted along with small pieces of charred root/stem (including one possible fragment of heather (*Ericaceae*) stem).

Discussion

Early Neolithic Pit Assemblages

The four assemblages were reasonably uniform in composition, although pit (605), sample <48>, did not include hazel nutshell fragments. There is little to indicate the precise nature of the material although, given the context, it is suggested that the remains may be derived from 'domestic' hearth waste. If so, this could indicate that the local population were still (at least in part) reliant on foodstuffs gathered from the wild. The material is largely typical of assemblages of Neolithic date, with numerous parallels both within East Anglia (e.g. Valley Belt, Trowse (Murphy 2000) and Over, Cambridgeshire (Fryer 2019)) and throughout lowland Britain (Moffett et al. 1989).

Middle Bronze Age–Earlier Iron Age

Given the extent and the apparent complexity of the ditched enclosure system, plant macrofossils are extremely scarce and give few firm indications of how the enclosures functioned on a day to day basis. A total of three possible cereal grain fragments and one wheat glume base are recorded from the Middle Bronze Age enigmatic burnt stone deposits within intervention (361) of the enclosure in Area A. The assemblage from enclosure ditch section (642) contained the only two weed seeds recorded from the site. However, the density of material recorded is too low to allow any further interpretation of the features.

Individual samples were also taken from a component post-hole of Gate Structure 1 (sample <33>) and from a further component post-hole of Fence-line 1 (sample <26>). Both contained cereals and charcoal/charred wood.

In summary, this site is somewhat unusual as it contains significant archaeological features, but little in the way of corroborative plant macrofossil evidence to assist with any interpretation. The similar site at Ormesby St Michael, Norfolk (Gilmour and Mortimer 2012) also recorded a large enclosure of Middle Bronze Age date, which possibly belonged to an extensive system of fields and stock enclosures. Similarly, excavations at Thorney near Peterborough (Pickstone and Mortimer 2011) and Trumpington, Cambridge (Phillips and Mortimer 2012) recorded extensive and complex enclosure systems contemporary with those from Swaffham. In all instances, anthropogenic detritus is scarce (including plant macrofossil assemblages) and it is difficult to ascertain through this medium precisely how such enclosures were used.

Pollen

FRANCES GREEN

Three pollen samples were analysed from the excavation. One sample was from monolith <09>, taken from Middle Bronze Age deposits in ditch segment (361). Samples <44> and <45> were from the fills of two Early Neolithic pits, (605) and (597) respectively (part of Pit Cluster 1).

Methodology: Pollen Preparation

Method 1

Pollen sample <44> was prepared by QUEST (University of Reading) using a heavy liquid separation technique. Sub-samples were extracted from the selected sequences/contexts by standard laboratory procedures (see Branch et al. 2005). The pollen is extracted as follows: (1) sampling a standard volume or weight of sediment; (2) deflocculation of the sample in 1 per cent Sodium pyrophosphate; (3) sieving of the sample to remove coarse mineral and organic fractions (>125 µ); (4) acetolysis; (5) removal of finer minerogenic fraction using Sodium polytungstate (specific gravity of 2.0 g/cm^3) (Munsterman and Kerstholt 1996; Nakagawa et al. 1998); (6) mounting of the sample in glycerol jelly. One tablet of lycopodium marker spores (Stockmarr 1971) was added at the acetolysis stage. Each stage of the procedure was preceded and followed by thorough sample cleaning in filtered distilled water. Quality control was maintained by periodic checking of residues and assembling sample batches from various depths to test for systematic laboratory effects.

Method 2

The two other pollen samples (<9> and <45>) were prepared using Hydrofluoric acid (Durham University). The pollen preparations followed Moore et al. (1991). One ml of sediment was desegregated by boiling in 10 per cent potassium hydroxide for five to ten minutes and passed through a 180 µm sieve. Silicates were removed by heating at 80°C in concentrated hydrofluoric acid (HF), the high silt and sand content requiring prolonged treatment. Acetolysis was performed to remove cellulose. The remaining material was stained, mixed with silicone fluid and mounted on slides using large square coverslips. A count of a minimum of 200–300 land pollen types per sample was attempted under a magnification of x400 and x1000. Pollen identifications were assisted by reference to Moore et al. (1991) and Andrew (1984), with fungal spores referred to Van Hoeve and Hendrikse (1998). Diagrams were drawn using TILIA (Grimm 1993) with most data expressed as a percentage of total land pollen (tlp, i.e., trees, shrubs and terrestrial herbs). Note that aquatics, spores and unidentified are, for example, expressed as a percentage of tlp + aquatics. Algae and fungi are expressed independently as a percentage of tlp (i.e. not including algae or fungi in the final sum). Plant nomenclature follows Fitter et al. (1985) with minor amendments from Stace (1997).

Results

Sedimentology

Sample <44> was subsampled from a bagged sample of deposit (598). Fill (598) of Early Neolithic pit (597) was a mid-greyish brown silty sand. Sample <45> was subsampled from a bagged sample of sediment from deposit (606), the fill of Early Neolithic pit (605). The overall character of the sediment was a dry mid-brown slightly silty sand with rare to occasional sub-rounded/sub-angular flint up to 20 mm in diameter. Monolith <09> was cleaned and recorded using the scheme of Tröels-Smith (1955).

Pollen

EARLY NEOLITHIC PIT (597), SAMPLE <44> (APPENDIX A, FIGURE 32, TABLE 11)

Fill (598) of pit (597) was a mid-greyish brown silty sand. A 6.2 g sub-sample from <44> was processed from this deposit and seventy-nine hard won land pollen grains were recorded from four coverslips. The pollen was well-preserved but sparse and the counts insufficient to be statistically valid, but an approximation of the local environment is possible.

Micro-charcoal was abundant. The fruiting bodies of soil mycorrhizal fungi *Glomus* were found at moderately abundant frequencies (14 per cent tlp + fungal remains). Mycorrhizal fungi *Glomus* are found in aerobic bioactive soils (Bagyaraj and Varma 1995), illustrating the backfill of the pit was derived from a bioactive soil. Fungal spores of both *Sordiaracae* (5 per cent tlp + fungal spores) and *Cercophora* (5 per cent tlp + fungal spores) were moderately common. Both are coprophilous fungi, living on herbivore dung and are indicators that grazing animals were present.

Tree pollen contributed 25 per cent of all land pollen (tlp). Alder (*Alnus*) (14 per cent tlp) was the most abundant with lesser amounts of pine (*Pinus*) (8 per cent tlp) and oak (*Quercus*) (4 per cent tlp). Pollen from shrubs was a minor component (3 per cent tlp) with small amounts of hazel (*Corylus*) (1 per cent tlp) and willow (*Salix*) (3 per cent tlp). Non-tree pollen predominated with grasses (*Poaceae*) the most important (39 per cent tlp) with *Taraxacum*-type (dandelion family) (27 per cent tlp) also an important component of this community. *Rumex acetosella* type (cf. sheep's sorrel) pollen was also present. A single grain of pollen from an interesting herb – *Circaea* sp. cf. *Circaea lutetiana* (enchanter's nightshade) – was identified. Enchanter's nightshade is a plant of shady and semi shady areas and is found in damp woods or borders of woods and hedge banks. In this location it suggests it is a relic of recently cleared woodland or that it grew under a canopy of alder mixed with perhaps oak close to the site.

EARLY NEOLITHIC PIT (605), SAMPLE <45> (APPENDIX A, FIGURE 33, TABLE 11)

Four coverslips were counted and a total of seventy-five grains of land pollen were recorded from the mid-brown silty sand of deposit (606) from pit (605). The pollen was moderately well preserved but limited in diversity, with only nine land pollen types identified. The predominance of dandelion family (*Taraxacum*-type) (36 per cent tlp), a pollen type which is resistant to corrosion hints that some of the assemblage may be lost. However, other more delicate pollen grains have survived and do not appear to be thin or lacy as expected if corrosion was a significant factor in the resultant assemblage.

The count of seventy-five pollen grains is not sufficient to be statistically valid but as in the pollen results from adjacent pit (597) allows a broad understanding of the possible landscape at the time the pit was infilling. The slide contained moderate amounts of opaque material which was likely to be finely divided charcoal. The presence of *Glomus*-type bodies (mycorrhizal fungi), which are found within aerobic bioactive soils (Bagyaraj and Varma 1995) indicates this was in part at least a bioactive soil. The presence of undifferentiated fungal spores of the *Sordariaceae*, a group associated with the dung of herbivores, probably indicate the presence of grazing animals.

The assemblage was dominated by non-tree pollen (*c.*70 per cent tlp) in particular those of grasses (*Poaceae*) (29 per cent) and dandelion family (*Taraxacum*–type) (36 per cent tlp). Pollen from other weedy taxa of open grassy and disturbed places included *Plantago* media/major (hoary plantain, common plantain) and probably *Ranunculus* sp. (buttercup family). Tree pollen contributed 17 per cent (tlp). Alder (*Alnus*) 16 per cent (tlp) was the most frequent tree pollen and is typical of wetter soils. Willow (*Salix*) was also found at low frequencies (1 per cent tlp) and would have grown in a similar environment. Hazel (*Corylus*) was present at low levels (3 per cent tlp). Pollen from forest trees was restricted to a single grain of oak (*Quercus*). Local heath is indicated by 11 per cent tlp of heather (*Calluna*). Spores of Pteridophytes including *Polypodium* and bracken (*Pteridium*) are resistant spores and are frequently overrepresented when part of the assemblage is lost due to corrosion. These two groups of ferns are however, consistently typical of a dry, open environment and are not overrepresented compared with other taxa and therefore likely to represent part of the local open and dry vegetation.

MIDDLE BRONZE AGE ENCLOSURE DITCH (361), SAMPLE <09> (APPENDIX A, FIGURE 34, TABLE 11)

Monolith <09> sampled the lower sediments of ditch (361). The single pollen sample assessed was from 28 to 29 cm from the top of the monolith and sampled lower ditch fill (376). Deposit (376) was a mid- to light yellow, brown clay-rich silt with frequent small 2–4 mm clasts of flint and chalk. The pollen preparation was full of extraneous silt and very little organic material. Occasional charcoal was recorded but the most striking presence was that of the 375 foraminifera tests counted in the four coverslips. These are derived from the underlying chalky till which was in turn partly derived from the underlying chalk where foraminifera are abundant. The presence of foraminifera from the chalky till implies an alkaline environment in the ditch which is not favourable to pollen preservation.

The pollen was found in low but countable numbers with eighty-eight total land pollen (tlp) being counted. The pollen was not noted as being especially corroded, but resistant pollen and spore types were commonly encountered; *Taraxacum* (dandelion family) (42 per cent tlp), *Compositae* (daisy family) (20 per cent) and also *Pteridium* (bracken). However, a very likely assemblage of trees, shrubs and herbs with a high proportion of *Poaceae* (grasses) (25 per cent tlp) suggests that largely this is a real reflection of the former environment although some of the more fragile pollen types may well have been lost in the alkaline environment of the enclosure ditch. Tree pollen only contributed 10 per cent tlp which illustrates a lack of local woodland with perhaps *Alnus* (alder) (8 per cent tlp) and *Corylus* (hazel) (1 per cent tlp) growing in the wetter or more open land and *Pinus* (pine) (1 per cent tlp) and *Quercus* (oak) (1 per cent tlp) growing at some distance or as isolated trees.

Discussion

Early Neolithic Pit Cluster 1

Pit (597) was one of the earliest in the sequence of eleven intercutting pits of Early Neolithic date, and pit (605) was stratigraphically in the middle. Evidence from Kilverstone (Norfolk) (Garrow et al. 2006) suggests that within a pit cluster each pit was dug and filled over a short period.

Despite the restrictive counts, pollen from pit (597) illustrates that the landscape, at least the immediate area of the pit, was open and grassy and that dandelion-type plants grew on the grazed (suggested by *Coprophilous* fungal spores) and disturbed ground. *Taraxacum*-type (dandelion-type) pollen is a robust and often differentially preserved pollen type and can therefore be overrepresented. Differential preservation may have enhanced its frequency, but the lack of corrosion observed in the other pollen grains suggests this has not had a huge impact on the overall interpretation. The animals which grazed the land may have been domesticated or wild.

Trees were largely cleared with either sparse scattered forest trees (pine and oak) or the woodland was at some distance. More locally, wetter areas supported alder which was the most important tree, as well as willow. Fine charcoal was found in abundance in the pollen preparation and a few larger fragments of charcoal were identified as large diameter oak and alder (see charcoal report, this volume). Oak was the most common charcoal type, albeit from a very small sample size, which corroborates the presence of oak in the landscape. It is tempting to think the charcoal was derived from continued clearance of the local woodland. A single pollen grain of enchanter's nightshade (*Circaea* sp.) may be a relic of recently cleared woodland or that the plant grew under a canopy of alder mixed with perhaps oak close to the site and local grass pollen production swamped the tree pollen record.

At a later date within the Early Neolithic, pit (605) was excavated and cut into an earlier cascade of pits. The pollen record from pit (605) suggests a predominantly dry (bracken and possibly *Polypodium*), grassy, weedy (dandelion-type), relatively open landscape. Tree pollen was present but was less frequent than in the earlier pit (605) and suggest trees and shrubs such as alder, willow and hazel were growing in wetter or marginal parts of the landscape. However, low frequencies of pollen from forest trees such as oak suggest oak woodland grew at some distance. Unlike the record from pit (605) pine pollen was not recorded and given its prolific pollen production and ability to be transported long distances, its absence indicates pine was probably not present locally.

There was no suggestion that the pollen within pit (605) was derived from cut flowers or plants intentionally placed within the pit, therefore the pollen assemblage must reflect the plants growing near the pit as it infilled. There was also no indication of arable activity. Comparison of the non-pollen assemblage (fungal spores) between pit (597) and pit (605) shows there was continued grazing of the site by herbivores, possibly domesticated.

High frequencies of dandelion-type (*Taraxacum*-type) pollen was a feature in common with the earlier pit (605), a plant which may have grown on the disturbed soil associated with the pit-digging activity. Local heathland is envisaged surrounding pit (605) where heather grew on dry and mostly treeless soils. The presence of heather strongly suggests a dry and treeless area in close proximity as *Calluna* pollen does not travel far. The development of acid heathland is likely to be the direct result of woodland clearance (Behre 1988) which may have increased in the interval between the excavation of pits (597) and (605).

Middle Bronze Age Enclosure Ditch (361)

Pollen evidence indicates the soils at the base of Middle Bronze Age ditch section (361) accumulated within a grassy almost treeless environment (with less tree cover than in Early Neolithic pit (605)), with alder growing perhaps in areas of wetter soil.

Areas of disturbed ground supporting weedy vegetation of the dandelion and daisy family were suggested by the pollen results. The presence of bracken may also suggest dry and possibly overused land.

No direct evidence of pastoral activity was identified, however, soil micromorphology of these same soils strongly suggests the grassy landscape was indeed grazed, although whether the animals were held within the enclosure or outside was not determined. Soil micromorphology also indicates that the lower soil was composed of two distinct soils which were physically mixed together, either by mesofauna (worms, etc.) or by human/animal trample. One distinct but unproven possibility is that within a grazing environment, animals would have entered the ditch for water and in doing so trampled the underlying sediments.

Conclusion

There is a similarity between the pollen from all features despite being widely separated in time. It appears there was significant woodland clearance in the Early Neolithic with at least local tree cover significantly lost and that the loss of all woodland, including that on wetter soils, continued further in the Middle Bronze Age. It is likely that a grazed, dry grassland with impoverished soils were dominant in both periods with local heathland periodically developed.

Diatoms

FRANCES GREEN

A single diatom sample was assessed from the lowest sediment (131) of Middle Bronze Age enclosure ditch intervention (129). If diatoms were present it would suggest the ditch contained permanent water when open.

Methodology

Diatom samples were prepared by boiling 2 cm^3 of sediment in 10 per cent Hydrogen peroxide for seven hours or until all the organic material had disappeared, and then mounting the sample in Naphrax. Routine counting under x1000 magnification attempted to count 200 frustules.

Results and Discussion

Despite two coverslips being scanned, no diatoms were encountered in the analysis of the sample of sandy clay silt deposit (131). A *Cocconies*-like frustule was observed, but this was too uncertain to include. Fifty-two foraminifera were recorded, derived from the underlying chalky till, which was itself in part derived from the Cretaceous chalk bedrock. The absence of diatoms strongly suggests that the part of the enclosure ditch around segment (129) did not contain permanent water when open. However, it is also possible due to the conditions of the burial environment that diatoms may not have been preserved.

Charcoal

FRANCES GREEN

Charcoal was recovered from features during processing of samples for the retrieval of plant macrofossil assemblages and bagged separately. This charcoal was studied to assess the range of tree species found within Early Neolithic, Middle Bronze Age and earlier Iron Age features.

Methodology

A selection of charcoal for identification was made randomly from the bagged samples. A sample size of twelve to twenty hand-picked fragments was attempted, but in most cases less than ten fragments were selected as fewer intact examples were present. Fragment sizes varied between 1 × 2 mm and 18 × 10 mm, with most falling in the 4 mm size fraction. No sieves were used. Charcoal fragments were broken along three main planes of identification (transverse (TS), tangential longitudinal (TLS) and radial (RS) sections). The sections were examined using a Meiji x40 stereomicroscope and as no higher magnification was available the identifications are limited by this. Identifications were made with reference to Schweingruber (1990) and Hather (2000).

For each charcoal fragment the number of rings counted and an example of the ring width was recorded together with an estimate of vitrification. A full catalogue from the analysis can be found in the archive (White 2020).

Results

Anatomical observations of 177 fragments of charcoal from twenty-four contexts identified six taxa *Quercus* (oak), *Corylus/Alnus* (hazel/alder), cf. *Betula* (cf. birch), *Salix/Populus* (willow/poplar) and cf. *Prunus/Maloideae* (e.g. blackthorn). The transverse section (TS) was the most diagnostic section available given the restraints of the magnification. Where identifications are not clear a 'cf.' is placed before the name, for example, cf. *Betula*. *Quercus* and *Fraxinus* are confidently separated from other wood types. In this study alder (*Alnus glutinosa*) and hazel (*Corylus avellana*) fragments could not be separated and are recorded as *Corylus/Alnus*. *Salix* and *Populus* are indistinguishable under any magnification so are recorded as *Salix/Populus*.

Charcoal was occasionally identified as cf. *Prunus/Maloideae* due to difficulty of distinguishing between the taxa in this group of *Rosaceae*. This group includes *Prunus* sp. e.g. *Prunus spinosa* (sloe/blackthorn), *Prunus padus* (bird cherry), *Prunus avium* (wild cherry) or *Prunus domestica var. insititia* (bullace), *Sorbus* (whitebeam) *Crataegus* (hawthorn), *Malus* (apples) and *Pyrus* (pears). Unidentified charcoal was not included in the context summaries presented in Figures 27 and 28 as it has no paleo-environmental value. The results are presented by period and calculated as a percentage of the total of all identifiable charcoal (n) for each context.

Early Neolithic Pits

Early Neolithic pit (597), fill (598), and pit (629), fill (630) contained few identifiable charcoal fragments. *Quercus* was the only wood type in pit (629) and *Corylus/Alnus* the most important (85 per cent total charcoal counted) in pit (597), with a smaller amount (15 per cent) of *Quercus*. All the wood had weak curvature suggesting mature wood with a larger diameter used rather than small branches and twigs.

Middle Bronze Age

Enclosure Ditch (361)

The charcoal from four deposits (376; 383; 385; 388) of enclosure ditch section (361) were analysed. Contexts (383), (385) and (388) were distinct burnt flint and charcoal-rich deposits (tips) in the upper levels of ditch (361). These deposits are suspected to derive from the levelling of an adjacent possible 'burnt mound' of Middle Bronze Age date. Deposit (376) was from closer to the base of the ditch and is probably contemporary with the original function of enclosure ditch (361). Charcoal in deposit (376) was dominated by relatively fast-grown oak (85 per cent) from mature (larger) timbers. *Corylus/Alnus* made up the rest of the charcoal which was from smaller wood.

The lower of the three sampled burnt flint and charcoal tips, (383), contained frequent charcoal of which the twenty-one fragments analysed indicated the presence of a range of species: *Quercus* (38 per cent), *Corylus/Alnus* (29 per cent), *Prunus/Maloideae* (14 per cent) and uniquely within the ditch fills, *Fraxinus* (ash) (19 per cent). These fragments showed a range of curvatures across all species, suggesting both large wood and twigs were used.

Stratigraphically above (383) was another thin layer of burnt flint and charcoal, deposit (385), which contained a similar amount of *Corylus/Alnus* (27 per cent) to the underlying deposit, but no *Fraxinus*. Other trees included *Quercus* (9 per cent), *Salix/Populus* (18 per cent) cf. *Betula* (18 per cent) and raised levels of *Prunus/Maloideae* (27 per cent) were also found. All of which were from moderate to smaller diameter wood.

Charcoal from the largest burnt flint and charcoal deposit (388), infilling the upper part of the ditch, contained a limited range of species. *Corylus/Alnus* was the most abundant (78 per cent) with lesser amounts of *Quercus* (11 per cent) and *Prunus/Maloideae* (11 per cent). The majority of the charcoal was from larger wood which contrasts with the smaller twiggy wood of the lower deposit. A single sample was taken from deposit (632). All the charcoal was *Prunus/Maloideae* and was from predominantly large diameter wood.

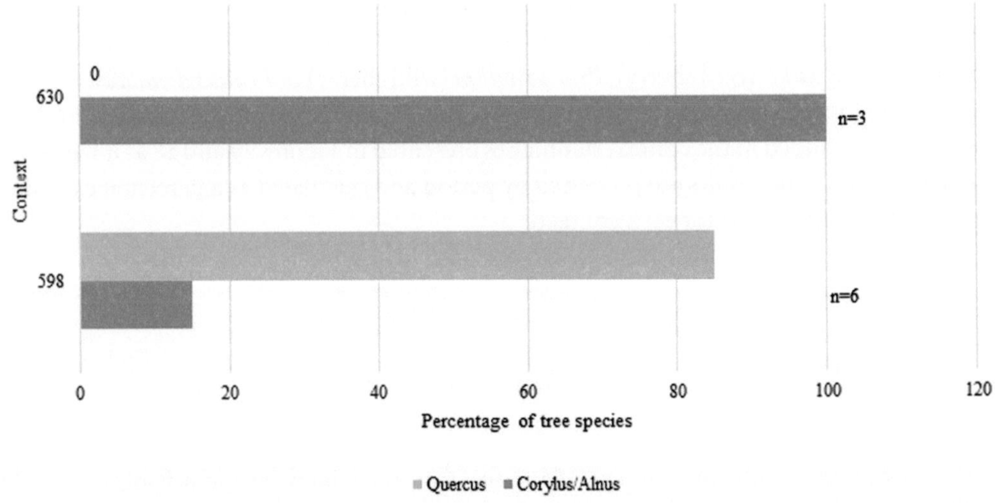

Figure 27. Charcoal frequency by species in Early Neolithic features. © c/o NPS Archaeology.

Enclosure Ditch (458)

Two deposits from ditch terminus (458), (460) and (462), contained small amounts of charcoal (n=6, n=4). *Quercus* was the most common in both samples with *Corylus/Alnus* making up the remainder in (460) and *Prunus/Maloideae* in (462). Both small and larger wood diameters were present.

Enclosure Ditch (550)

Charcoal from fill (545) of ditch segment (550) was from open scrubby woodland and hedgerow shrubs/trees. *Corylus/Alnus* was the most common (50 per cent), with lesser amounts of cf. *Betula* (17 per cent) and *Prunus/Maloideae* (33 per cent).

Enclosure Ditch (671)

All twelve charcoal fragments from ditch terminus (671), deposit (674), were of *Quercus*. The fragments sampled were all large with up to eight tree rings and ring widths up to 1.5 mm wide. The fragments had little to no curvature and were derived from large diameter wood.

Enclosure Ditch (704)

Charcoal from deposit (719) within corner section (704) of the enclosure ditch was very similar to that found in enclosure ditch intervention (550). Of the four fragments, two were *Corylus/Alnus*, one *Prunus/Maloideae* and one *Salix/Populus*.

Enclosure Ditch (776)

Two charcoal fragments were identified as *Corylus/Alnus*.

Pit (265)

Charcoal was recovered from three fills of clay-lined pit (265) (probably associated with the 'burnt mound' activity). *Corylus/Alnus* (hazel/alder) was the most frequent charcoal type in fill (266) and (296), making up 100 per cent of all charcoal in deposit (266) and 90 per cent in (296). *Quercus* (oak) made up the rest of the assemblage (10 per cent) in (296). The *Corylus/Alnus* charcoal in (266) were all from smaller branches and twigs, and the charcoal in (268) was also from small diameter wood. Deposit (298) differed in that it contained a wider range of species, none of which were dominant. These included twigs and slightly larger diameter wood of *Corylus/Alnus* (20 per cent), *Prunus/Maloideae* (e.g. blackthorn) (30 per cent), *Quercus* (10 per cent) and *Populus* (poplar) (10 per cent).

Middle Bronze Age–Earlier Iron Age Post-Holes

Quercus was the most frequent charcoal type in post-holes (164) (Structure 1) and (439) (Gate Structure 1). Post-hole (164) also contained *Fraxinus*. In most examples the *Quercus* charcoal was from large diameter wood and other taxa such as *Corylus/Alnus*, *Prunus/Maloideae* and *Salix/*

Populus was from a mix of large and smaller branches and twigs. Post-hole (220) (Fence-line/ Structure 3) contained an isolated fragment of *Salix/Populus* from mature wood. Post-hole (770) (Post-hole Group 3) contained low frequencies of identifiable charcoal and that included a mix of *Salix/Populus*, *Prunus/Maloideae* and *Quercus*.

Discussion

Early Neolithic Pits

The two pits (597) and (629) contained such low charcoal counts that no real conclusion can be drawn except that larger oak and hazel/alder was growing locally and was being burnt. The probable diameter of the wood used suggests there is a possibility it was derived from bonfires and clearance.

Middle Bronze Age Enclosure Ditch

The significance of species variation across the infilling of ditch (361) needs consideration in view of the small sample size. However, it does appear that during the early use of this ditch, fuel was mainly derived from mature oak with a smaller contribution of hazel or alder roundwood, perhaps as kindling.

Oak charcoal is not as important in the upper fills of ditch (361) compared with those at the base, and an upwards pattern of reduction in the utilisation of forest trees is hinted at. A range of both forest and hedgerow resources seems to have been used in the burnt flint deposits. Perhaps

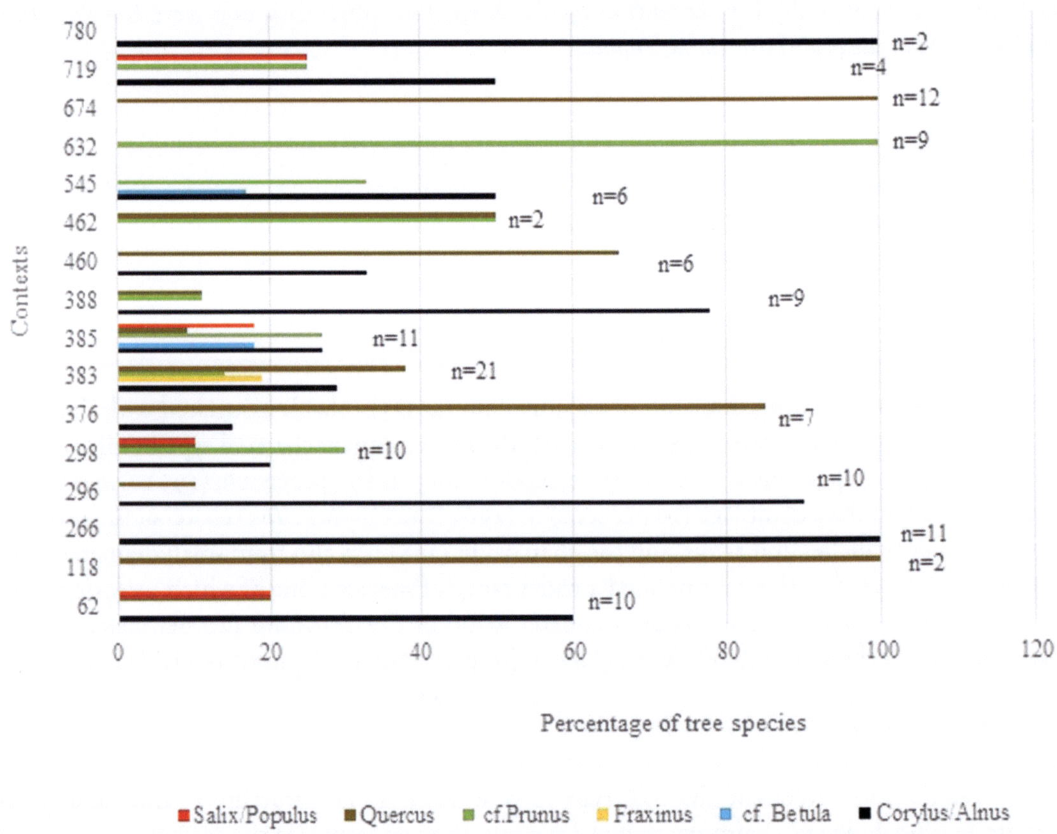

Figure 28. Charcoal frequency by species in Middle Bronze Age features. © c/o NPS Archaeology.

significantly, there is an overall decline in forest trees being used in the uppermost deposit. Both ash and oak are found in the lowest burnt flint tip but in the uppermost tip hazel/alder is most common, perhaps hinting oak and ash (both excellent fuel) were not as abundant in the landscape. Pollen evidence however shows that during the early phase of the ditch the immediate landscape was largely treeless apart from perhaps smaller areas of alder and hazel, suggesting wood was collected for fuel at some distance from the enclosure. A single sample from deposit (632) contains only charcoal from mature hedgerow bushes and small trees, possibly even fruit trees.

The charcoal in the three tips of burnt flint deposits contained different tree species. For example, ash and willow are found in separate deposits. While some of this variation may be explained by the small sample sizes, it also may be due to charcoal being derived from different sources. The size of utilised wood also varied, with that from upper deposits derived from larger boughs. This suggests the burnt flint was from another small-scale source other than a (redeposited) burnt mound. However, it is also possible that material of a possible burnt mound accumulated in a stratified way with fuel sources varying over time and that as the ditch was backfilled and the possible 'burnt mound' levelled, different layers were redeposited.

A clay-lined pit (265) thought to be associated with the burnt mound activity, also contained charcoal. Charcoal of hazel/alder in deposits (266) and (298) from pit (265) were very similar to the upper fill (388) of Middle Bronze Age ditch (361). Although in the pit, the charcoal was from twigs whereas charcoal in the ditch fills was from larger bough and trunk wood.

Charcoal from burnt mounds often appears to be indiscriminately derived from the local landscape with alder common in wetter areas. For example, mixed but *Corylus/Alnus* dominated fuel charcoal from the Bronze Age burnt mound at Northwold, Norfolk (Crowson 2004) and Feltwell Anchor (Bates and Wiltshire 2000), which may be in part derived from the seasonal clearance of vegetation at these sites. The potential example from Swan's Nest however appeared to use both forest and scrubby woodland resources in varying amounts over time.

Charcoal from other enclosure ditch sections (550; 704; 780) was in main derived from hedgerow and scrubby hazel/alder wood. The origin is unknown and may be from local hearths or bonfires. Terminus (458) contained charcoal from both larger timber from woodland together with branches and twigs from hedgerow and open scrub. The charcoal associated with animal bone is likely to be from dumped hearth waste. Terminus (674) of the outer enclosure ditch was atypical of any of the other charcoal samples in that it contained exclusively oak charcoal, many in large fragments. The charcoal was entirely from mature large diameter timber and it is possible it was derived from a single burnt timber.

Conclusion

In summary, this study has shown oak was locally present from the Early Neolithic to the Middle Bronze Age. Ash was also found in the Middle Bronze Age local woodland. The branches and sometimes larger wood of hedgerow shrubs were commonly utilised throughout the life of the enclosure ditch, as were scrubby woodland trees such as hazel/alder.

Detailed interpretation of results is constrained by the small sample sizes, but some overall conclusions can be made. Interestingly, when the enclosure was first open it appears large oak timber and smaller hazel/alder was probably selected for cooking fires, and no hedgerow trees were utilised. Did potential hedging therefore come later in the life of the enclosure, or does this reflect the point at which hedges at the enclosure become mature?

No conclusive evidence has been produced to support the inference that burnt flint fills of the upper part of enclosure ditch (361) were derived from a 'burnt mound' in the vicinity. Although undoubtedly the great volumes of thermally shattered flint were produced by the rapid cooling of heated flint in water; whether this material was stockpiled in a 'burnt mound' first is not clear. However, a clay-lined pit possibly associated with this activity contained a charcoal assemblage with similarities to the upper fills of enclosure ditch (361). Despite the tiny sample size such a similarity may suggest the upper fills of the ditch were not from the bottom of a burnt mound but was actually the last material to be fired. A tentative but unproven possibility is that the charcoal and burnt flint was not stockpiled first but simply tipped into the enclosure ditch as it was created.

Charcoal associated with the burnt flint in the enclosure ditch seems to suggest that wood selection was ad-hoc and a variety of local wood resources were utilised from forest, hedgerow and scrubby woodland. However, there is a suggestion that the differing tree species present between burnt flint layers may in fact represent various separate events, either from more than one area of stockpiled burnt flint or the tips themselves represent different heating events.

Micromorphology

ROWENA Y BANERJEA

This section summarises the findings arising out of the micromorphological analysis undertaken by Quaternary Scientific (QUEST), University of Reading. The analysed sample was taken from the Middle Bronze Age enclosure to establish the type of activities undertaken at the site, and to explore if they were of a pastoral or occupational nature. The deposit that was sampled was silt unit (376) of enclosure ditch segment (361), which contained Middle Bronze Age pottery and animal bone. Above this deposit, multiple tipped-in charcoal-rich burnt flint deposits were present, which have been interpreted as having derived from 'burnt mound' activity.

Methodology

One micromorphology sample was prepared from a sub-sample collected from monolith <09>, 24–32 cm from the top. The sampling procedure followed the University of Reading standard protocol for thin section preparation. The sample was oven-dried to remove all moisture and then impregnated with epoxy resin while under vacuum. The impregnated sample was then left overnight so that the resin could enter all of the pores. The sample was then placed in an oven to dry for 18 hours at 70°C before it was clamped and cut to create a 1 cm slice through the sample. The surface of the 1 cm slice was flattened and polished by grinding on the BROT. The prepared surface of the 1 cm slice was then mounted onto a frosted slide and left to cure. This was followed by cutting off the excess sample, so the sample was down to a thickness of 1–2 mm. The mounted sample was ground down to approximately 100 µm in thickness using the BROT. The 100 µm section was then lapped on a Logitech LP30 precision lapping machine to the standard geological thickness of 30 µm.

Micromorphological investigation was carried out using a Leica DMLP polarising microscope at magnifications of x40–x400 under Plane Polarised Light (PPL), Crossed Polarised Light (XPL), and where appropriate Oblique Incident Light (OIL). Thin-section description was conducted using the identification and quantification criteria set out by Bullock et al. (1985) and Stoops (2003), with reference to Courty et al. (1989) for the related distribution and microstructure, Mackenzie and Adams (1994) and Mackenzie and Guilford (1980) for rock and mineral identification, and Stoops et al. (2018) for further identification of post-depositional alterations. Tables of results use the descriptions, inclusions and interpretations format similar to Bullock et al. (1985) and Simpson (1998) (Appendix B).

Micromorphology enables the following properties to be examined at magnifications of x40–x400 under PPL, XPL and OIL: thickness, bedding, particle size, sorting, coarse: fine ratio, composition of the fine material, groundmass, colour, related distribution, microstructure, orientation and distribution of inclusions, the shape of inclusions and finally the inclusions to be identified and quantified. In addition, post-depositional alterations can be identified and quantified such as: effects on the microstructure by mesofaunal bioturbation and cracking due to shrink-swell of clays or trampling; translocation of clays and iron; chemical alteration such as the neoformation of minerals such as vivianite and manganese; organic staining as a result of decayed plant material; and excremental pedofeatures such as insect casts and earthworm granules.

Results and Interpretation of the Micromorphology

Micromorphology descriptions for each deposit are recorded in Table 13 (Appendix B), the frequency and types of inclusions within these deposits are recorded in Table 14 (Appendix B) and the abundance of post-depositional alterations and pedofeatures within the deposits are recorded in Table 12 (Appendix B). To determine the deposit type classification, each deposit was grouped using the following diagnostic sedimentary attributes and inclusions which provide crucial information concerning the origin of inclusions, transportation mechanisms of particles and the deposition processes. To ascertain the origin of sediment components, descriptions were made of particle size, shape and the composition of the coarse and fine fraction, particularly the frequency of rock, minerals and anthropogenic inclusions (Table 14, Appendix B). The depositional events are characterised by the following sedimentary attributes: sorting, related distribution, orientation and distribution of the inclusions (Table 13, Appendix B) and bedding structure (Table 14, Appendix B).

Understanding the formation processes for deposits is crucial to interpreting the depositional pathways of rock fragments and minerals, any anthropogenic debris such as charred wood and artefacts, and other types of plant remains and microfossils (Banerjea et al. 2020). Analysis of post-depositional features provides crucial information concerning the effects of weathering, preservation conditions (Stoops et al. 2018) and stratigraphic integrity of the deposit (Canti 2003, 2007; Macphail 1994). Micromorphological analysis has identified that deposit (376) is formed from two types of microfabric (MF) that are mixed, MF1 and MF2.

Microfabric Type 1 (MF1)

MF1 (Table 13, Appendix B) has a sandy clay particle size and is unsorted. The fine material is calcareous and yellowish-brown/greyish brown, PPL, and silvery/golden (3rd order interference colours), XPL, in colour. The inclusions are embedded within the fine material and are unoriented and unrelated, and random and unreferred in their distribution, which represents a haphazard organisation. The spongey microstructure may have formed as a result of mesofaunal activity and vughs may represent collapsed air pockets (Stoops et al. 2018).

The inclusions (Table 14, Appendix B) are predominantly geological comprising fragments of angular and sub-angular flints, 20 per cent, sub-rounded chalk fragments, 20 per cent, quartz, 30 per cent, calcite, 5 per cent and muscovite, 2 per cent. Herbivore dung was identified, 10 per cent, containing calcareous faecal spherulites, 2 per cent, which form in the guts of some ruminants (Canti and Brochier 2017).

Microfabric Type 2 (MF2)

MF2 (Table 13, Appendix B) has a sandy clay loam particle size and is unsorted. The fine material is calcareous with iron and mid-brown, PPL, and orange brown/greyish brown, XPL, in colour. The related distribution of the inclusions is linked and coated, whereby the fine material forms bridges between and coats the coarse inclusions, and embedded and coated, whereby the coarse components are embedded within the fine material. They are unoriented and unrelated, and random and unreferred in their distribution, which represents a haphazard organisation. The spongey microstructure may have formed as a result of mesofaunal activity (Stoops et al. 2018).

The inclusions are predominantly geological comprising fragments of angular and sub-angular flints, 15 per cent, and some that might be fired, 5 per cent, sub-angular and sub-rounded chalk fragments, 30 per cent, quartz, 30 per cent, calcite, 5 per cent, plagioclase, 2 per cent, and microcline, 2 per cent. Fragments of charred wood were identified, 5 per cent, and calcareous faecal spherulites, 2 per cent, which form in the guts of some ruminants (Canti and Brochier 2017) occurred in areas of organic staining. Two types of sediment aggregates were identified: rounded, sub-angular and sub-rounded fragments of an iron micropan, 10 per cent, and aggregates of a sub-rounded silty clay loam fabric, 5 per cent.

Post-depositional Alterations

Both MFs show effects of weathering, organic decay and reworking of the stratigraphic sequence by bioturbation (Table 12, Appendix B). Iron nodules and organic staining, arising from the decay of organic materials, such as dung, occurs in both MF types; however, iron nodules are more abundant in MF2 than MF1, 5–10 per cent and 10–20 per cent respectively. Iron nodules are redoximorphic pedofeatures (Vepraskas et al. 2018) and probably relate to a fluctuating water level within the ditch. The spongey microstructure in both MF types can be indicative of surface soil horizons (Stoops et al. 2018) whereby biological activity increases porosity.

Discussion

One micromorphology sample was collected from deposit (376) of the Middle Bronze Age enclosure to establish the nature of the activities within and/or surrounding the enclosure and to see if the activity is representative of occupation, the corralling of animals or both.

Summary of Formation Processes

Micromorphological analysis shows that the area where this sample was collected from in deposit (376) is formed from two types of sediment, MF1 and MF2, which have distinctive sediment attributes. MF1 and MF2 are mixed, or churned in a 'puddled-like' way. This could have occurred pre-deposition whereby two types of sediment were either dumped into the ditch and again reworked by post-depositional mesofaunal bioturbation, or churned as a result of post-depositional trampling in wet conditions (Rentzel et al. 2017), and/or mesofaunal bioturbation within the ditch which caused the rotation of the sediments and infilling voids with sediment (Kooistra and Pulleman 2018).

The slightly greater abundance of iron nodules within MF2 may suggest that the latter is a more likely scenario – that the sediments were deposited separately and then reworked by post-depositional processes. The greater abundance of iron nodules in MF2 could indicate that it underwent the redoximorphic effects of a fluctuating water levels within the ditch while deposited as a separate layer – fragments of an iron micropan would support this. However, it could also be possible that the greater amount of iron nodules formed locally because of decaying organic material and a fluctuating water table.

Depositional Pathways of Inclusions

Artefacts, herbivore dung and charred wood within deposit (376) are likely to have been deposited within at least two different stratigraphic units or be contained within two different materials that were redeposited within the ditch; the former is the most plausible explanation.

Conclusions

The location in deposit (376), within monolith <09> where this sample was collected from, is formed from two types of sediment, MF1 and MF2, which have distinctive sediment attributes. These two MFs are thought to have been originally two different stratigraphic units which have since been mixed by people or animals trampling in the ditch and/or by mesofaunal activity. Consequently, any material culture or herbivore dung remains that were recovered from or identified within deposit (376) represent at least two different depositional events.

CHAPTER 5

Swan's Nest Environs Aerial Imagery Assessment

CHRIS COX

Introduction

An aerial imagery assessment of land within the Swan's Nest environs was carried out by Air Photo Services, centred on UK National Grid Reference TF 8247 0755, coordinates 582470, 307550. This section considers all aerial and satellite imagery and Lidar data which were readily available for archaeological examination in October 2020. The objective of this assessment was to:

- Provide information on the location and nature of buried and upstanding archaeological features;
- Provide a landscape context for the complex Middle Bronze Age enclosure at Swan's Nest; and
- Investigate the possibility of associated sites or field systems which may be present in the immediate area.

Archaeological and Natural Features from Aerial Imagery

In suitably cultivated soils, sub-surface features including archaeological ditches, banks, pits, walls or foundations, may be recorded from the air in different ways in different seasons. In spring and summer these may show through their effect on grass and crops growing above them. Upstanding features which may survive in un-ploughed grassland are also best recorded in winter months when vegetation is sparse, and the low angle of the sun helps pick out slight differences of height and slope.

Limitations of Aerial Photographic Data

Aerial photographic evidence is limited by seasonal, agricultural, meteorological and environmental factors which affect the extent to which either buried or upstanding archaeological features can be detected from the air. The visibility of archaeological features may differ from year to year and is dependent on the type of crop or land use, prevailing weather conditions and the levels of moisture in the soil over the grass or crop growing season. Differences in the intensity and angle of the light also assist greatly in seeing the nuances of slightly upstanding earthwork features. Individual photographs often record only a small percentage of the actual extent of buried or upstanding features and a wide range of photographs taken over a long timescale may be needed to reveal the extent of buried features visible from the air.

The Study Area

In this area of Norfolk, the chalk substrate is well drained, and crops respond readily to differences in the depth and consistency of the topsoil, over areas where buried bur and embanked features are present. This effect also applies to anomalies in the consistency of the substrate. Aerial images in this region show widespread marks in crops over large areas of 'patterned ground' which are caused by these geological patterns and anomalies in the chalk (Stephens 1990: 121–4). These patterns are particularly visible in some areas of the study area and are easily discernible from crop marks caused by archaeological features which are more regular and obviously anthropogenic.

Methodology

All images were interpreted and mapped at a level compatible with a 1:2500 scale base map. The photographs were downloaded or clipped at high resolution from internet-based mosaics, and closely examined in detail on screen and as layers within the online geoportals.

Images were digitally rectified to an OS base map using the QGIS rectification tool. This was done to remove perspective distortion and ensure correct rectification of aerial photographs to the OS map projection EPSG 27700 British National Grid (Scollar 2002). Images were also interpreted and rectified to OS map bases and used in accordance with observations made by Scollar and Palmer (2008).

In all transformations prepared for this assessment the mean stated mismatches were less than ± 2.5 m. The rectified files were set as background layers in QGIS where features were interpreted and drawn over the rectified photographs. Layers from the final drawing have been used to prepare Figure 29 and are contained digitally in the archive for import to GIS in ESRI Shapefile format.

Data Sources

Available aerial and satellite images provide considerable heritage data, and were derived from all the timelines at google.com/earth (1999, 2006, 2018, 2019 and 2020) and aerial imagery at bing.com/maps and magic.defra.gov.uk/magicmap, along with mosaicked vertical aerial photographs taken in 1946 (RAF) and 1988 (Geonex) displayed at historic-maps.norfolk.gov.uk. Vertical and oblique aerial photographs held as prints at the HE Archive in Swindon and the NHER archive in Gressenhall were not consulted due to their closure during the Covid-19 pandemic.

The Aerofilms collection at britainfromabove.org.uk contains aerial photos of Swaffham town centre dating from 1928, but unfortunately none of these covered the study area outside of the town. The Cambridge University Collection of Aerial Photographs (CUCAP) was not open for consultation during the timescale of this assessment and has not been available for some time as is currently being digitised for archiving by the university. The CUCAP database, at cambridgeairphotos.com/search indicates that there are no specialist obliques which contain crop marked evidence, only those listed as recording the geological patterns in crops which are prevalent in this area. However, close interpretation may possibly reveal some archaeological features. The images used for this assessment are listed in the archive (White 2020).

Lidar data were downloaded from the Environment Agency website and were visualised and interpreted alongside historic and modern aerial and satellite imagery. These data were gathered in 2017 at a 1.00 m resolution for the NLP.

Results

The assessment identified eighteen sites or areas of archaeological interest within the study area and two further sites within the immediate environs which have been included for completeness of data. The sites which have been identified by this assessment are illustrated in Figure 29 and discussed below. A full detailed catalogue of the sites identified can be found in the archive (White 2020).

At the site of Swan's Nest (Site 18), few changes were identified across time from that of the 1885 First Edition Ordnance Survey Map. The only changes detected were with the field immediately north of the site, with a single residential development constructed in its eastern half adjoining Watton Road. A subsequent vertical aerial photograph taken in 1988 shows the incorporation of the small parcel of land left over after the above-mentioned development, with the Area A field to its south. No changes to the boundaries or to the size of the two fields investigated (Area A and Area B) have taken place between the taking of this aerial photograph in 1988 and the completion of the 2018–2019 archaeological excavation.

Discussion

The environs of the prehistoric enclosure at Swan's Nest have been heavily ploughed, but it may be expected that remains of buried ditches, banks, previous settlement enclosures and possible funerary sites will be visible as crop marks in suitable conditions. This assessment has shown that, even in the absence of consultation of specialist oblique aerial photographs derived from archaeological survey, there are positive indications of the presence of pre-modern features in this area.

Site 1 is identified by a small length of crop marked ditch and a possible ditch, which may indicate outlying features to the south of Swan's Nest. Other features to the north-east of Swan's Nest comprise a series of crop marked ditches and likely enclosures and boundaries at Sites 7, 9 and 10. These sites were all identified from images displayed at Google Earth, most notably those taken during what is obviously a summer month (likely late June or early July) by Getmapping plc in 2006. Fragmentary remains of ditched enclosures and possible fields are recorded in this area. Site 5 may be the result of modern agricultural processes, but one element is curvilinear, and could possibly be a buried heritage asset. This interpretation is tenuous and remains to be tested by future research.

In the south part of the study area, Sites 15 and 16 are well-defined buried heritage assets, which again illustrates the archaeological features in comparison to the appearance of the underlying geological strata as crop marks. Site 15 is a circular ditched enclosure, which at 15 m diameter is a little small to be definitely identified as the ring ditch surrounding a Bronze Age funerary site, but may fall into this category. To its immediate north, Site 16 is a well-defined buried double-ditched trackway with likely associated settlement enclosure and ditches. Again, these sites show clearly as crop marks on images taken by Getmapping and displayed at the 2006 (dated 31/12, but likely summer) timeline at Google Earth. The area contains numerous possible pits and large swathes of 'patterned ground' cropmarks indicative of anomalies in the chalk substrate. The features here are likely to be more complex than initially shown on the aerial image. Lidar data indicates no microtopography and any remaining features will lie below the surface and are likely to be partially cut into the chalk through thin topsoil. Features are likely to extend to the south of the visible focus of the enclosure, track and ring ditch.

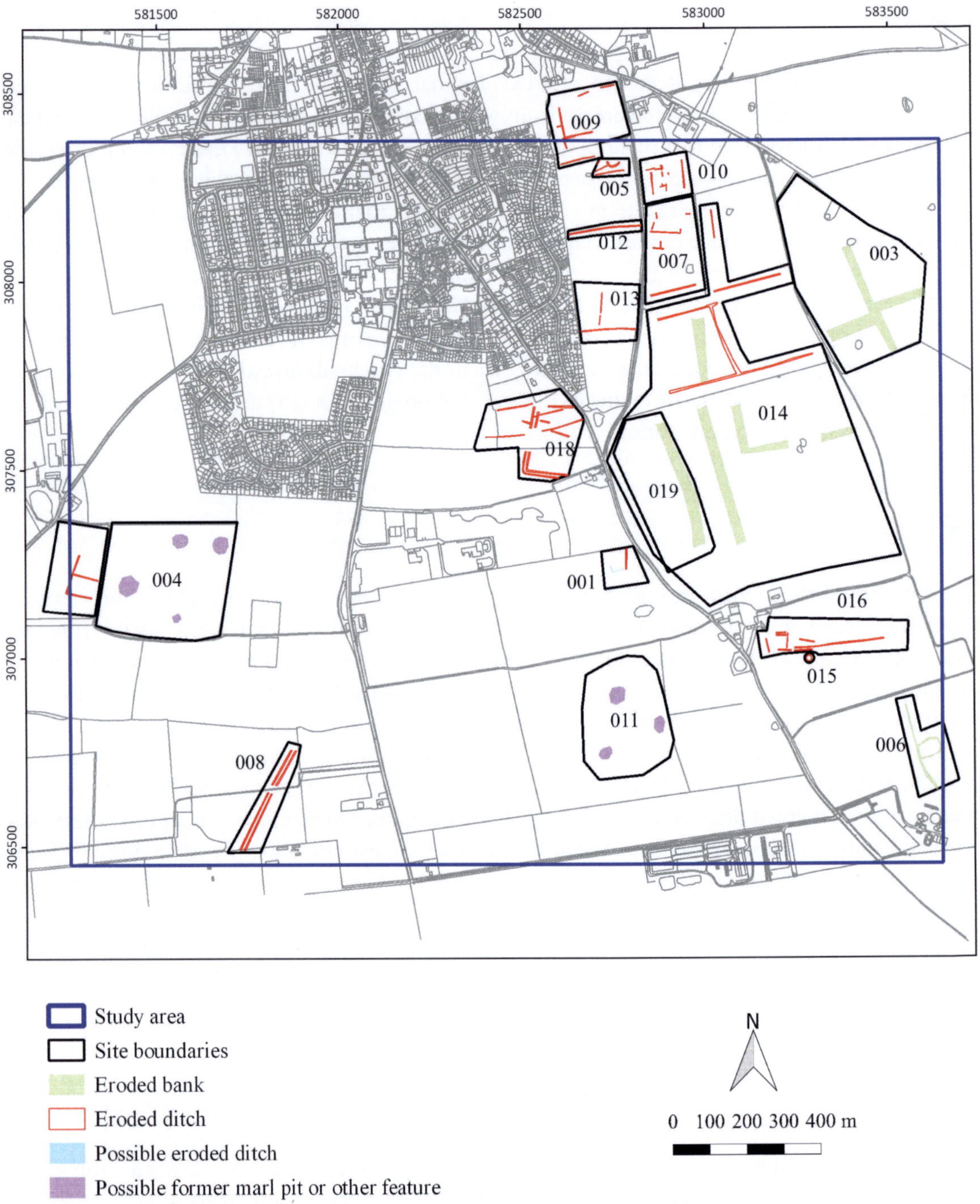

Figure 29. Identified sites within the study area. © Air Photo Services.

Site 17 is again visible at Google Earth, 2006, and is a crop marked partially visible ditched enclosure with straight sides and an internal division. While currently undated, this enclosure may form part of the pre-modern or prehistoric landscape, and either be a small focus of rural settlement or a stock penning site.

Outside of the immediate study area, Site 20 to the north is a well-defined crop marked D-shaped ditched enclosure which is cut by a modern field boundary. The site is not mapped in detail, but is visible on the 1946 RAF mosaic of aerial photographs displayed at historic-maps.norfolk.gov.uk. It is included to indicate the continuity of likely prehistoric features in this area.

Site 2, if viewed in isolation, may be easily mistaken for the eroded ring ditch surrounding a Bronze Age round barrow. The circular cropmark is caused by the sub-surface remains of part of the former Second World War military airfield at North Pickenham, which has now been removed and returned to arable cultivation. This is attested by the 1946 RAF image, which shows the site as an extant feature.

Site 6 is a former coppice and boundary. The small wood was extant in 1946, subsequently removed and is now visible as a light-toned soil mark. Further light-toned amorphous areas of different soil are visible at Site 19 to the east of Swan's Nest. Previous observations of similar areas on chalk geologies in Norfolk (Cox 2017) and Hertfordshire (Cox 2013) indicate that these areas may have potential for buried prehistoric enclosures and settlement sites. In areas with a similar appearance where a wide range of specialist oblique aerial photos and NMP mapping data are available, buried enclosures, tracks and pits have been recorded as crop marks over similar features. Lidar data indicates slight topographic anomalies which point to possible marl pits, but the area is heavily eroded.

Post-medieval sites in the study area comprise the eroded remains of enclosure field boundaries (Sites 3 and 14) and likely former very eroded marl pits or possible other cut features at Sites 4 and 11. Site 8 is an undated, possibly modern, double-ditched track which now lies in woodland and is visible on both earlier aerial imagery at Google Earth and the 2017 Lidar data DTM.

Conclusion

The study area contains well-defined evidence for likely prehistoric or pre-modern sites which may have had a landscape association with the main enclosure at Swan's Nest. However, no distinct evidence for a contemporary field system surrounding the enclosure was identified. Marks in crops and soil from the Swan's Nest environs indicate a range of ditched enclosures, a track, some possible field boundaries and significantly, a ring ditch which may have been a funerary site.

CHAPTER 6

Discussion

Mesolithic

Worked flints typologically dating to the Mesolithic period were recovered as residual artefacts from features of later dates. Prior to the discovery of this assemblage, the only evidence for Mesolithic activity in the area was the recovery of a single microlith found during excavations *c.*150 m to the south-west of the site (Sleap and Savage 2016).

The flint assemblage from Swan's Nest attests to an appreciable hunter-gatherer presence in the area during the Mesolithic, with the finding of a tranchet adze from the site adding to the corpus of such tools known from west Norfolk. Murphy (1984) has highlighted how during the Mesolithic, the heavily forested terraces of the Breckland area appear to have been capable of supporting substantive semi-nomadic, seasonally mobile populations, which were able to efficiently exploit the available natural resources. The artefacts recovered from Swan's Nest appear to support this view.

Early Neolithic

Evidence of Early Neolithic activity at the site was represented by a cluster of eleven intercutting pits and four post-holes. This activity appears to date from the middle Early Neolithic, indicated by the recovery of Mildenhall Ware ceramics and a radiocarbon date of 3645–3521 cal. BC (Beta – 552730, 95.4 per cent), which came from a cotyledon fragment of a *Fabaceae* recovered from Pit Cluster 1. A large assemblage of residually deposited struck flint of typologically Early Neolithic date was also recovered from the upper fills of the Middle Bronze Age–earlier Iron Age enclosure ditch.

The Nature of the Occupation

The group of pits at Swan's Nest is characteristic of the 'developed' pit clusters recorded at other archaeological sites in the region. In particular, significant similarities can be drawn between the remains from Swan's Nest with those recorded at Kilverstone, Norfolk (Garrow et al. 2006). At Kilverstone, 236 steep-sided pits in discrete clusters were excavated, with average diameters of 0.7 m and average depths of 0.23 m (Garrow et al. 2006). A programme of radiocarbon dating established that the pit-digging activity at Kilverstone lasted approximately 150 years from the mid-thirty-seventh to mid-thirty-fifth centuries BC (Whittle et al. 2011: 333).

The Swan's Nest pits had similar forms to those identified at Kilverston, with steep-sided profiles, average diameters of 0.65 m and average depths of 0.28 m. Further commonality can be seen with the contents of the pits, with dark humic fills identified at both sites, along with similar assemblages of material culture. At Swan's Nest, pottery, struck flint, hazelnuts and rare fragments of animal bone were recovered from the Early Neolithic pit cluster; a strikingly similar assemblage to what was retrieved *c.*25 km to the south-east at Kilverstone (Garrow et al. 2006). Further Early Neolithic pit clusters akin to the example identified at Swan's Nest have also been excavated at Hurst Fen, Mildenhall in Suffolk (Clark et al. 1960), Broome Heath, Ditchingham in Norfolk (Wainwright 1972) and at Spong Hill, North Elmham, also in Norfolk (Hills and Lucy 2013).

The artefacts recovered from Swan's Nest indicate that a standard range of 'domestic' activities typical of the period were likely carried out at the site. No evidence was identified to suggest the occurrence of specialised activities; however, flint scrapers were a significant component of the artefact assemblage. Minor evidence for food processing was identified from the zooarchaeological and plant macrosfossil analyses, along with an abundance of charcoal suggesting that hearths were a feature of the occupation.

In this region, the identification of such 'developed' pit clusters is considered to be indicative of the presence of Early Neolithic settlement (Garrow et al. 2005, 2006; Anderson-Whymark 2011; Last forthcoming). Consequently, it is appropriate to consider that a settlement of sorts existed a Swan's Nest during the middle Early Neolithic. It is important to highlight that settlement in this period cannot be understood solely through sedentary frameworks, with evidence conversely suggesting that during the Neolithic, populations were fairly mobile (Thomas 1999).

Despite the transient nature of Early Neolithic society as a whole, based on the available evidence, the Swan's Nest 'settlement' is still best defined as having been of a comparatively small-scale, with only a single pit cluster present, in contrast to the sites outlined above which are characterised by the occurrence of numerous separate pit clusters.

Based on the results of the analysis carried out upon the remains from Kilverstone, it is likely that the Swan's Nest occupation represents a single episode of seasonal activity by a small to moderate-sized group, which was not repeated. Through the refitting of artefacts, at Kilverstone it was established that admixing of contemporaneous artefact assemblages across pits in the same cluster was a consistent phenomenon, but contrastingly, admixing of contemporary assemblages was not seen across separate pit clusters (Garrow et al. 2005). Although several scenarios have been put forward to explain this pattern, Garrow et al. (2005) favoured an interpretation which viewed a single pit clusters as being dug by a small group of people across the duration of a visit, with subsequent (?seasonal) visits creating further discrete pits clusters.

However, at the same time, caution must be taken when attempting to define the duration of the Early Neolithic activity at Swan's Nest. Firstly, none of the artefacts from Pit Cluster 1 were able to be refitted, and thus the pits within this cluster could not be categorically proven to be from the same episode of occupation. Garrow et al. (2005) has highlighted that through experimental archaeology, infilled pit clusters are recognisable for several years from the surface, which emphasises the potential for pits to be sequentially added to earlier clusters during later visits. Secondly, that the pit cluster at Swan's Nest existed in isolation cannot be proven beyond reasonable doubt. The large enclosure ditches present across the site could have conceivably truncated away other 'developed' pit clusters, and further examples may be present immediately beyond the edge of excavation, which were not identified.

A notable component of the sequentially excavated pit cluster at Swan's Nest was that it formed a 'hook' or 'question-mark' shape in plan. Such a pattern is not unique to the site, with similar such formations identified at Kilverstone. A variety of suggestions have been proposed to explain the creation of these formations, although most have shortcomings. Garrow et al. (2005, 2006) has discussed their potential development around trees, buildings or midden heaps, but has added that their formation/patterns may simply result from a specifically 'managed' process.

A distinct aspect of the Early Neolithic activity at Swan's Nest was the presence of a group of four post-holes (Post-hole Group 1) to the east of the pit cluster. This is not a component frequently seen at similar such sites discussed above, where post-holes are rare. It is difficult to interpret this group, with the component features being widely spread-out and appearing to have an almost wide rhomboidal arrangement. It is possible that they may be structural in nature and relate to dwellings or similar, however no likeness could be identified with any of the known Neolithic structures outlined by Darvill (1996). It is possible that Post-hole Group 1 may derive from an animal corral, fence-line or may have had a similar yet identified function.

Markers of Place?

Interpretations of Early Neolithic pit clusters have ranged from the practical and mundane, to the metaphysical and metaphorical. Thomas (1999) has stated that these features are largely unsuitable to have acted as grain silos (following Clark 1960), with most interpretations conversely favouring that the creation of pit clusters and the deposition of artefacts within them results from the interplay between complex social dynamics related to ritualised deposition and the marking of place (Thomas 1999; Chapman 2000; Garrow et al. 2005; Chadwick 2012). Despite such pits appearing to have been dug simply to receive deposits/artefacts (Pollard 2001), as Garrow et al. (2005) states, there appears to have been a distinct lack of 'structure' to this process, with little consistency in the type and quantity of objects deposited, changes in horizontal and vertical artefact distributions throughout pit fills and the varying angles at which artefacts are deposited (indicating they were tipped rather than 'placed'). Such depositional characteristics were also observed amongst the remains from the Swan's Nest cluster.

Although these processes are still only partially understood (Garrow et al. 2005), it is likely that the Swan's Nest pit cluster was formed through the same mechanisms. Unfortunately, the data retrieved from Swan's Nest was unable to shed further light behind the intricacies and purpose of these processes.

The Environmental Evidence

Analysis of pollen from the Early Neolithic pit-digging activity at the site established that the contemporary environment consisted of grazed and partially disturbed rough open grassland. Locally, willow and alder grew in wetter areas, with deciduous woodland present at some distance. Interestingly, no evidence for arable cultivation was identified – a feature also apparent in the Middle Bronze Age sample.

Both wild and domestic animals likely grazed the land surrounding the site, as evidenced by the presence of fungal spores. The presence of domestic cattle remains from Pit Cluster 1, although not categorical, represent a strong indication that domestic cattle were probably kept by the inhabitants of Swan's Nest during the Early Neolithic. Analysis of charcoal also confirmed the utilisation of oak and alder by the Swan's Nest community during this period.

Interestingly, a second pollen sample was analysed from a stratigraphically later pit in the sequence, which contained less tree pollen compared to the stratigraphically earlier sample. As has been discussed, interpretations of 'developed' pit clusters favour that the pits within a cluster were dug and backfilled throughout a single episode of occupation. However, it has been outlined how such a temporal interpretation cannot be confidently applied to the Swan's Nest occupation, with no evidence of artefact refitting. Consequently, this variation in tree pollen could represent the sequential digging of pits in the same cluster across many years, as well as provide direct evidence for the continued clearance of woodland in the landscape during the Early Neolithic. However, it remains entirely possible that this variation in tree pollen is simply the result of non-biotic taphonomic processes. Ultimately the small-scale nature of the Early Neolithic remains at the site limits the degree to which we can garner information about the human communities and their environment during this period.

Middle Bronze Age

The Middle Bronze Age activity was the most significant identified at Swan's Nest. Remains dating to this period consisted of a large rectangular 'Ormesby-type' double-ditched enclosure (approximately 2.1 ha in size), along with two probable structures, three probable fence-lines, an entrance structure and numerous discrete post-holes and pits. Evidence for the occurrence of 'burnt-mound' activity was also identified at the site during this period.

Activity Chronology

The primary indication of the Middle Bronze Age date for the casting of enclosure and the associated initial phase of activity at the site came from the recovery of Deverel-Rimbury style ceramics from the primary and secondary fills of the enclosure ditches. This typological date indication was supported by four radiocarbon dates.

The earliest radiocarbon date came from a fragment of large mammal bone from the base of the enclosure ditch. Although this fragment did not derive from an articulated specimen, with potential that it could have been residually deposited, it provided a plausible approximate date for establishment of the enclosure, around 1501–1323 cal. BC (Beta – 552731, 95.4 per cent). A short period of time after the casting of the enclosure, 'burnt mound' activity commenced at the site. The date of this activity was established through the radiocarbon dating of two charcoal fragments, which returned dates of 1497–1305 cal. BC (Beta – 552729, 95.4 per cent) and 1450–1291 cal. BC (Beta – 552728, 95.4 per cent). A further radiocarbon date was retrieved from an articulated juvenile cattle specimen found in the upper fill of the enclosure ditch. This returned a late Middle Bronze Age/early Late Bronze Age date of 1386–1128 cal. BC (Beta – 552732, 95.4 per cent). This suggests that the site was probably continually in some form of use for at least a century, possibly much longer, by which point parts of the enclosure ditches had become considerably infilled.

The date of the activity at Swan's Nest is comparable to several sites in the region. At the site of Ormesby St Michael, Norfolk, the casting of the enclosure was dated to approximately 1440–1260 cal. BC (SUERC-29974, 95 per cent), with the structural remains inside the enclosure dated to 1420–1210 cal. BC (SUERC-29975, 95 per cent) (Gilmour and Mortimer 2012). This may suggest that the setting-out of the Swan's Nest enclosure predates the type-site of the monument class to which it appears to belong. At Brigg's Farm, Thorney in Cambridgeshire, a similar enclosure and associated activity was dated by an extensive series of radiocarbon dates to between *c.*1400–1250 BC (Pickstone and Mortimer 2011). Equally, at Clay Farm, Trumpington in Cambridgeshire a series of radiocarbon dates established that the enclosures identified during excavations dated from *c.*1450–1260 BC, with the associated settlement dating from *c.*1520–1120 BC (Phillips and Mortimer 2012).

It appears that the enclosures outlined above were established and in use at a broadly similar time during the Middle Bronze Age, and that Swan's Nest was contemporary with a number of other enclosures in the region.

Environment and Setting of the Enclosure

Analysis of environmental remains indicates that the Swan's Nest enclosure was situated in a grassy, almost treeless landscape, with some alder growing around wetter areas. Little to no evidence of arable cultivation was identified at the site or in the surrounding vicinity. Pastoral activity was likely dominant in the environs, with the site's grassland setting providing a suitable environment for rearing and keeping domestic livestock, particularly cattle, sheep and horses.

Evidence indicating that animals were confined within the enclosure came from pollen and micromorphological analysis, which identified the presence of animal dung and trampling within the central enclosure ditch. Zooarchaeological remains have emphasised the importance of cattle at the site, with evidence for on-site breeding identified through the recovery of neonatal bones. Zooarchaeological and molluscan evidence additionally demonstrated that mature and well-established hedges were likely a key component of the site – a factor not identified from the sub-surface archaeological features alone. The stockpiling of wood also likely occurred at the site, which was identified through this same medium.

The saturation of the enclosure ditches was a key feature during the excavation, with many of the *c.*1.5 m deep ditch slots becoming entirely filled with ground and rainwater within a few days of being excavated. It was originally considered that the water-holding potential of the enclosure ditch may be a key feature behind its situation in the landscape, through providing an important source of water for the cattle rearing and 'burnt mound' activities known to have been undertaken at the site (see below). However, molluscan and diatom evidence showed that the enclosure ditches did not contain permanent water. However, there may have been some seasonal holding of water in the enclosure ditches across the winter months (of which evidence has not survived), or alternatively, the potential of the enclosure to hold water in prehistory may have been different than that of today, due to various potential changes to the local landscape and environment.

Enclosure Development and ?Field System

Based on the date of the enclosure, its form (including its overall size, the double-ditched arrangement, the size of the ditches and its segmentation into two approximately equal portions) and nature of the material culture assemblages recovered, it is appropriate to consider this site as belonging to the 'Ormesby-type' enclosure monument class (after Gilmour et al. 2014).

The emergence of such sites in the eastern region is contemporary with the first significant division of the landscape of lowland Southern Britain, which also occurs alongside a cultural shift from the Early Bronze Age, the appearance of expansive field systems and significant sedentary settlements. Although the precise purpose of 'Ormesby-type' enclosures is only partially understood, they seem to develop within predating or contemporary ditched field systems, where they appear to become a central element and subsequently often give rise to settlement.

At Ormesby St Michael, prior to the casting of the enclosure, a Middle Bronze Age co-axial field system was laid out. The Middle Bronze Age enclosure was later incorporated into this field system, on the same alignment (Gilmour et al. 2014). A similar such pattern of development was also identified at Clay Farm (Phillips and Mortimer 2012), Thorney (Pickstone and Mortimer 2011) and at Langtoft in Lincolnshire (Hutton 2008).

However, a predating or contemporary field system was not identified from the sub-surface archaeology recorded at Swan's Nest. A single ditch (Ditch 6) shared an alignment with the enclosure, and although this may represent an element of such a field system, this feature could not be confidently dated. In order to help understand the situation of the Swan's Nest enclosure, aerial photographic analysis and lidar work was commissioned as part of this project, with the primary aim of identifying potential evidence for a surrounding field system or similar. No such evidence was found from the consulted sources, however some (probably unrelated) significant prehistoric buried remains were discovered *c.*600 m to the south-east of the site. Despite the analysis of aerial photographic data failing to identify any evidence for prehistoric field systems in the vicinity of Swan's Nest, the discovery of a Middle Bronze Age ditch at land west of Brandon Road has been significant in confirming that at least a partially parcelled and organised landscape existed beyond the confines of the Swan's Nest enclosure. This Middle Bronze Age ditch identified *c.*750 m south-west of Swan's Nest has been radiocarbon dated to 1503–1423 cal. BC (SUERC – 96528, 95.4 per cent) (Blackbourn 2021), close to the date recovered from the base of enclosure ditch slot (103), calculated at 1501–1323 cal. BC (Beta – 552731, 95.4 per cent). The 'L-shaped' ditch at land west of Brandon Road is considered by the excavators to also represent an enclosure, with a series of post-holes perhaps once part of an associated gate structure (*ibid.*: 14).

On balance, although much less defined and apparent than at other sites in the region, it is highly likely that the open grassy landscape surrounding Swan's Nest was partially parcelled-up during the

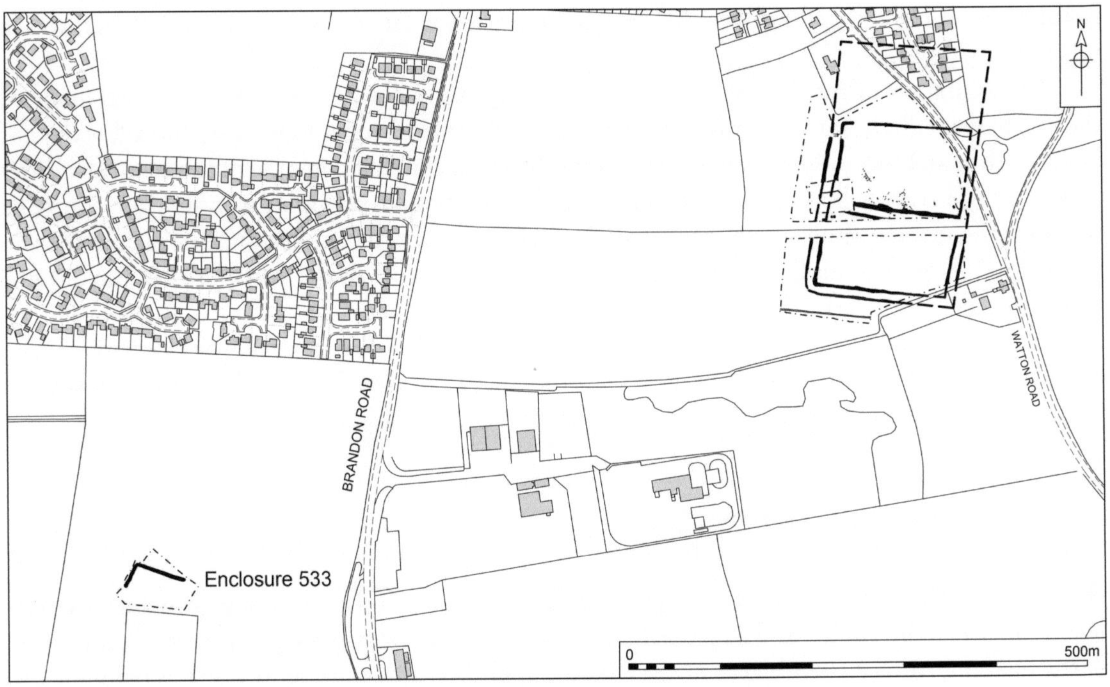

Figure 30. Possible projection of the Swan's Nest enclosure and enclosure 553 at land west of Brandon Road (after Blackbourn 2021). © Crown copyright and database rights 2021 Ordnance Survey 100019340.

Middle Bronze Age, with scattered enclosures and boundaries present throughout the area. Several suggestions could be proposed to account for the more elusive character of a field system or a parcelled landscape in this context, and one such scenario could be that the lack of Bronze Age ditches identified during excavations in the vicinity of Swan's Nest may not be a genuine reflection of their absence, but a result of such features having been incorrectly dated (see White 2020: 10–21) or having been undatable. Alternatively, it is also possible that surrounding field systems and enclosures may not have been predominately defined by ditches. Creating a ditched field system across the heavy clay soils may have been considered an intimidating or possibly unnecessary task. Fence-lines or hedgerows may have been preferentially used across the landscape; which would be unlikely to leave a significant archaeological footprint, if any. The field systems of Clay Farm (Phillips and Mortimer 2012), Brigg's Farm (Pickstone and Mortimer 2011), Langtoft (Hutton 2008), Ormesby St Michael (Gilmour and Mortimer 2012), Newark Road and Fourth Drove (Pryor 2001) are predominantly situated on softer peat, sands and gravels – in stark contrast to Swan's Nest.

It is also important to emphasise that the full extent of the Swan's Nest enclosure was not uncovered by this excavation, with evidence indicating that it continued further to the north. It is possible that it was in this area that the enclosure may have connected to a predating/contemporary Middle Bronze Age field system. A possible projection of the enclosure to the north has been presented in Figure 30 and a reconstruction illustration presented in Figure 31.

The Swan's Nest Enclosure: Purpose and Function

Such sites are considered to have played a role in pastoral management, with large banks and ditches constructed to contain livestock, potentially defending them from raiding or wild animals (Gilmour et al. 2014). The enclosure at Ormesby St Michael was situated in a low-land area surrounded by a wetland environment, with good access to rich grazing lands; further supporting the idea that these sites may have played a role in livestock husbandry (*ibid.*). The low-lying situation of the centre of the Swan's Nest enclosure, where water may have accumulated seasonally, along with its location on

Figure 31. Reconstruction illustration of the Swan's Nest enclosure during the Middle Bronze Age (Period 3) by Gary Collyer. © c/o NPS Archaeology.

the poorer quality clay-lands may additionally speak to its presence within a landscape primarily, if not exclusively, exploited by pastoral agricultural regimes. Swan's Nest is also located approximately 1.20 km to the north of the dry valley which possibly once formed the head of the River Gadder (now at Cockley Cley); a considerable nearby source of water and verdant pasture in prehistory.

As previously discussed, there is some evidence to suggest cattle breeding was being carried out at the site, and micromorphological analysis has confirmed the presence of animals having been contained within the enclosure. Cattle were also the most abundant species represented amongst the faunal remains recovered. At Clay Farm, the presence of cattle was confirmed through the recovery of dung beetles and parasites, along with the identification of stock pens within the enclosure (Phillips and Mortimer 2012). Consequently, the function of these enclosures within cattle husbandry regimes seems highly probable.

However, the scale of the banks and ditches that formed the Swan's Nest enclosure would seem to far exceed what would be necessary to contain large herds of cattle, a feature also recognised by Gilmour et al. (2014) with Ormesby St Michael and by Hill (1999) concerning similar Iron Age enclosures. Although the site may have had an important defensive function, its setting (surrounded by higher ground) arguably runs counter to this interpretation.

An alternative suggestion is that its monumental elaboration relates to the site's potential significant symbolic importance to the community that constructed it – an interpretation favoured for many such sites, particularly in the Iron Age (Hill 1999). As to whether this excessive construction represented a marker of place, status, ownership, had a ritual purpose or something else is, however, difficult to ascertain. Pryor (2001) has emphasised how similar such sites in the Fens would have likely acted as communal marketplaces for the exchange of livestock and a plethora of regular social gatherings.

Following from this, a distinguishing feature of the Swan's Nest enclosure was that its form was exceptionally regular, suggesting it was constructed in a precise, organised and managed fashion. When compared to contemporary enclosures in the region (outlined above), this defined and precisely organised arrangement is not typically observed. In many ways, it shares more similarities in form with the Late Iron Age viereckschanzen-type enclosures, such as at Fison Way in Norfolk (Gregory 1992; Davies 1999). The scale of the (excavated part of) the enclosure is also notable, with an approximate size of 2.1 ha, compared to Ormesby St Michael which is projected to have enclosed *c.*1.2 ha.

Pickstone and Mortimer (2011) suggested an average construction capacity of 8 m of ditch digging a day for an adult individual. With a total approximate length of 840 m of ditch forming the Swan's Nest enclosure, a very rough estimate is that the known extent of the Swan's Nest enclosure would have taken 105 person days to construct. If this was done by an individual farmer or small family group which could spare a day a week, the enclosure would have taken approximately two years to construct. A larger community or an extended family, would have been able to construct the enclosure considerably quicker. When compared to contemporary enclosures, Swan's Nest stands out as being a particularly substantial constructive feat. As to whether this had ramifications in relation to its comparative capacity or status within Middle Bronze Age society, is not clear.

The Structures and Fences

Several fence-lines and structures considered to date to the Middle Bronze Age were identified inside the enclosure in Area A. The fence-lines most likely blocked physical/visual/metaphysical access to different parts of the enclosure, or aligned internal routes, as might be the case for Fence-line/Structure 3. Unlike at Clay Farm, Trumpington, none of the fence-lines present at Swan's Nest could be confidently categorised as representative of stock pens.

Fence-line 1 was of particular interest in that it appeared to respect both the area of 'burnt mound' activity and the sloping parts along the northern edge of the enclosure ditch in this location, mimicking the form of the 'truncation'. Could this represent the screening-off of the 'burnt mound' activity due to ritual or metaphysical practices? Or did this fence-line simply physically restrict access by animals and/or people to what may have been a particular boggy part of the enclosure, or a defined activity area which needed to remain sectioned off?

Two structures were identified in the central southern part of Area A, although their function/purpose cannot be discerned from the available evidence. Similarities between Structure 1 and Structure 2 can be seen with Middle Bronze Age Structure 217 identified at Ormesby St Michael which is of a similar size and shape (Gilmour and Mortimer 2012). Such structures may represent shelters, houses or even sweat lodges associated with the 'burnt mound' activity discussed below. The Swan's Nest structures also appear to share similarities with the one identified at Chimney Farm, Witton in Norfolk, which was dated to the Middle–Late Bronze Age (Bates 2008).

As to whether these structures indicate the existence of occupation at Swan's Nest which could be defined as a 'settlement' is not clear. A relatively restricted domestic artefact assemblage was recovered from the site; however, such a feature has not prevented similar sites in the region being defined as 'settlements'. A distinct aspect of the Swan's Nest occupation however is the (probable) absence of the defined roundhouse structures recorded at the sites of Brigg's Farm (Pickstone and Mortimer 2011) and Langtoft (Hutton 2008).

The 'Burnt Mound' Activity

A series of deposits of concentrated burnt flint identified in the northernmost internal dividing ditch are thought to have originated from a 'burnt mound' originally situated close to this part of the enclosure. The identification of this industrial-scale activity on site is similar to Bronze Age activity recorded at Postwick, Norfolk (Crawley 2014), Northwold, Norfolk (Crowson 2004), Marham Park, Suffolk (Green 2018) and Feltwell Anchor, Norfolk (Bates and Wiltshire 2000).

At Swan's Nest, deposits of crushed/highly fragmented burnt flint in a charcoal and silt matrix had been tipped from the south into the enclosure ditch, from a narrow area between two linear features, where an up-cast clay with chalk bank appears to have been present. The confined nature of the space where this activity appears to have been concentrated is intriguing. As to why this location was seen as most appropriate for this activity is difficult to ascertain, although its low-lying position could have resulted in better access to ground/rainwater, or perhaps the raised platform created by the up-cast from the bank was seen as beneficial. Within this space, a clay-lined trough/pit was identified along with a post-hole, both of which appeared to be relate to the 'burnt mound' activity due to their fills containing heat-altered flint. Similar such component features are frequently identified at other 'burnt mound' sites in the region (Bates and Wiltshire 2000; Crawley 2014; Green 2018).

Flint was likely gathered from the superficial geological clay deposits at the site, within which large nodules are abundant. Despite this abundance, the same burnt flint material appears to have been used on numerous occasions, where it became fractured down into extremely small fragments. At the point at which it became unusably fractured, it appears to have been discarded. The charcoal analysis indicated that a variety of fuel sources were used in the combustion process, although poorer quality wood which was likely gathered from the immediate environment appears to have been favoured. Although there is no obvious stable water source at the site, the presence of two inland 'burnt mound' complexes at Marham Park, Bury St Edmunds demonstrates how such sites did not need to exist immediately alongside open natural water sources, with water most likely supplied through wells and rain at this site (Green 2018). It is entirely possible that a well was also present at Swan's Nest beneath one of the unstripped areas. Alternatively, a spring may have existed at the site in prehistory (Frances Green pers. comm.), or perhaps seasonal ground and rain water accumulations within the ditches (as witnessed during the excavation) were adequate enough sources to allow the 'burnt mound' activity to be carried out.

The burnt flint material does not appear to have been deposited into the ditch in a single episode. Between deposits of burnt flint, redeposited natural was present, likely from bank-slumping or deliberate backfilling. It is more likely that these deposits represent deliberate backfilling due to the repeated nature of their depositional pattern and the consistency of their make-up. As to why burnt flint deposits were 'capped' or concealed with deposits of clay is not clear. Indeed, as to why the burnt flint was interred into the enclosure ditch, as opposed to being left to form an upstanding mound, is also unclear. This element of the Swan's Nest evidence is particularly significant in light of interpretations frequently presented that view the mounds formed through these industrial-scale processes as having been particularly symbolic and special (see Bates and Wiltshire 2000; Crowson 2004).

The presence of silt washes between the deposits of burnt flint indicates that this depositional sequence was created over a moderate period of time, possibly suggesting that the 'burnt mound' activity was repeated. However, this is only a precise reflection of the rate of material deposition, not the combustion activities themselves.

A variety of interpretations have been put forward to explain such large-scale 'burnt mound' activity identified at archaeological sites. This has included suggestions of cooking, industrial (dyeing) or possibly a ritual/ceremonial function (Crowson 2004). The precise purpose of the 'burnt mound' activity at Swan's Nest is not clear, but most probably relates to one of the above suggestions.

Swan's Nest represents the first definitive example where defined, near 'industrial'-scale heating of flint has been identified within an 'Ormesby-type' enclosure. Such activity has previously been hinted at, with a trough-like pit similar those seen at other 'burnt mound' sites identified at Ormesby St Michael in association with a structure (Gilmour and Mortimer 2012). Significantly, as outlined above, this structure was similar to those identified at Swan's Nest, supporting the idea of a shared function for their interpretation. These structures could represent small dwellings or shelters where heating, cooking (Ó Drisceoil 1988) or industrial activities (Jeffery 1991; Crowson 2004) were carried out. It is also possible that such structures could have had a ritual function, possibly acting as prehistoric saunas (Barfield and Hodder 1987).

Late Bronze Age

A small assemblage of Late Bronze Age pottery was recovered from the tertiary deposits of the Middle Bronze Age enclosure ditches in two discrete concentrations. The presence of these sherds is not considered indicative of any considerable occupation at the enclosure during this period, but they are significant in confirming the continued sporadic use of the site through the Late Bronze Age, prior to a more concerted reuse of the enclosure during the earlier Iron Age. Although a hiatus in activity at the site likely occurred at some point during the Late Bronze Age, the presence of deposits containing ceramics from this period suggest that this pause in activity was not extensively prolonged. It is also possible that these sherds originate from a 'hybridised ceramic phase' (Percival: this volume), and thus may be broadly contemporary with the Deverel-Rimbury occupation.

Little evidence for arable cultivation was detected at the site and its surrounds during the Middle Bronze Age, with activity instead appearing to focus (partially) on the rearing and management of livestock, particularly cattle. It is possible that pastoral (grazing?) activities continued at the site to some degree sporadically throughout the Late Bronze Age.

The lack of general evidence for extensive Late Bronze Age activity at the site is of significance, and is a pattern also reflected at Ormesby St Michael (Gilmour and Mortimer 2012), Clay Farm (Phillips and Mortimer 2012) and Brigg's Farm (Pickstone and Mortimer 2011), where very little to no evidence of Late Bronze Age activity has been identified. This trend may reflect shifts in local communities' social affiliations and approaches to such enclosure 'monuments', or is perhaps a reflection of such sites becoming economically redundant, and possibly situated in areas of depleting agricultural productivity. It is possible that this deviation may correlate with the large-scale socio-economic changes occurring throughout the Late Bronze Age across Britain (Yates 2007; Darvill 2010).

Earlier Iron Age

The primary feature of the earlier Iron Age activity at the site was the reutilisation of the enclosure earthworks; the banks are suspected to have still been upstanding and the ditches partially open during this period. The main indicator of this reuse was the recovery of Post Deverel-Rimbury pottery (with decorative and formal traits) from the secondary and tertiary fills of the enclosure ditches.

The Enclosure: Continuity or Repurpose?

No clearing or recutting of the enclosure ditch was identified, with the casting of Linear 1 representing the only modification to the enclosure dating to this phase of occupation. This shallow linear ditch-like feature was respectfully wedged between the two Middle Bronze Age ditches in the south-west part of the enclosure. It is unclear as to what this feature represents or what its original function was. It is unlikely that it represents a linear quarry hollow, as the Middle Bronze Age bank which would have consisted of 'marl' up-cast, would had to of been cleared for its construction. One possible interpretation that can be proposed through analysis of the feature's form and dimensions, is that it arguably resembles how the Middle Bronze Age enclosure ditches would have looked when partially filled during the early–mid-first millennium BC. This would possibly indicate a direct attempt to elaborate and extend the enclosure in a way which was in keeping with its overall form. As to whether this extension had a mundane practical purpose, simply acting as a 'repair' to maintain the corralling and containing potential of the enclosure, or whether this modification had a more symbolic function, is not clear. The lack of enclosure ditch recutting however, which would have suitably achieved the purpose of the former suggestion, may add more weight to the latter.

If this feature does represent an elaboration of the enclosure during this period, this would imply that the activity at the site during the earlier Iron Age was more than simply transient casual pastoral reuse. This is further supported by the presence of Post-hole Group 3, which likely represents some sort of structure, to the east of Linear 1. Little evidence was recovered that helped to define the nature of the activity, but the presence of the pottery and animal bone suggests that a degree of domestic activity was undertaken at the site. The recovery of Iron Age loomweights from Post-hole Group 3 and the enclosure ditch during the trial trenching (Peachey 2016b) provide the only indications of craft production from the site, with casual small-scale textile manufacturing likely undertaken. The small size of the ceramics assemblage perhaps implies that occupation was either short-lived or was more ephemeral but temporally repeated. However, as previously outlined, this may conflict with the casting of a substantial linear elaboration to the enclosure. A greater understanding of the duration/temporality of the Iron Age occupation has been limited by the lack of suitably secure deposits on which to carry out radiocarbon dating – although such a programme would have likely been hindered by the 'Hallstatt plateau' (Hamilton et al. 2015).

As to whether the earlier Iron Age occupation at Swan's Nest was of a similar nature to that of the Middle Bronze Age, or was of an entirely different character, is also difficult to ascertain from the available evidence. However, if the original function of the enclosure was partially related to its environmental setting (as detailed above), it is likely that such intrinsic factors were present into the first millennium BC, resulting in the potential for the same economic functioning of the site to continue. A notable concentration of earlier Iron Age ceramics was retrieved from the enclosure ditches around the entrance/exit identified in the north-west of Area A, suggesting this was still very much in use during this later phase of occupation.

Although simply speculation, it is possible that some level of arable cultivation may have occurred at the site during the Iron Age. The tertiary fill sequence of the enclosure ditches appears to be mostly represented by leeched soils which had moved into the ditch. As to whether this occurred naturally or through the action of ards, is unable to be discerned; however, the notable deposition of Early Neolithic struck flint into the tertiary fills of the enclosure, suggests the possible disturbance of a partially sealed earlier activity spread during this period (see Garrow et al. 2005). Struck flint associated with the Early Neolithic occupation of the site was not found in abundance across the primary and secondary fills, implying the deposits which contained this material remained largely secure and undisturbed throughout the earlier life of the enclosure. It appears that during this final occupation and tertiary infilling of the enclosure ditches, this changed, with the commencement of wide-scale cultivation activity at the site representing a possible explanation for this. However, other explanations to account for such dynamics may also be likely.

Iron Age Reuse of 'Ormesby-type' Enclosures

Strikingly similar activity to that recorded at Swan's Nest was excavated at Clay Farm, Trumpington, Cambridgeshire, where Early Iron Age reoccupation of a Middle Bronze Age enclosure was identified (Phillips and Mortimer 2012). A key feature of this activity was the recovery of Early Iron Age pottery from the partially filled enclosure ditch, as was seen at Swan's Nest. The digging of large pits also occurred during this period at Clay Farm, and although refuse pits were not identified at Swan's Nest, Linear 1 was of a similar intrusive scale. At Brigg's Farm, Thorney in Cambridgeshire, two large ring gully structures typologically dated to the Middle to Later Iron Age were recorded, illustrating further that the Iron Age reoccupation of Middle Bronze Age enclosures appears to represent a distinct phenomenon in the region.

Iron Age Settlement Patterns in Norfolk

Hill (1999) has characterised the settlement landscape patterns of the Iron Age in Norfolk as 'spurgy'. This is in contrast to the 'spotty' settlement patterns which are prevalent throughout most of Britain, comprised of 'discrete, spatially isolated and bounded' entities, typified by 'Little Woodbury' style settlements consisting of small enclosed farmsteads (*ibid.*: 190). 'Spurgy' settlement however is distinctly unenclosed, ranging over large areas, with the domestic nuclei shifting over time (*ibid.*). It has been suggested that this settlement pattern in Norfolk was unlikely to have be dictated by environmental factors or that it represents the manifestation of a more 'peaceful' society (*ibid.*). Conversely, it is believed to result from complex social and symbolic factors; large wandering settlements may indicate societies whose land rights were more communally held and field systems less permanent and more open (*ibid.*).

Although the occupation at Swan's Nest can be articulated within a 'spurgy' understanding of the settlement landscape, with few dated features and few indicators of intense activity (implying a short-lived or ephemeral but repeated occupation); it also is situated within an enclosure. The attraction to the enclosure in this period and its reutilisation illustrates that enclosed occupations were not fundamentally at odds with settlement practices in this region. Additionally, little in the way of Iron

Age activity has been identified in the surrounding vicinity despite several large-scale excavations. Consequently, it is difficult to currently categorise the Swan's Nest evidence within a ranging 'spurgy' settlement pattern or a 'spotty' pattern.

Unlike the more 'developed' occupation identified at Clay Farm, Trumpington (Phillips and Mortimer 2012), the Swan's Nest occupation may have represented more ephemeral temporally repeated activity. One possible suggestion outlined above is that the site may have played a role in cattle husbandry regimes, possibly similar to in the Middle Bronze Age, being seasonally occupied at certain points of the year to exploit the surrounding grasslands. Pickstone and Mortimer (2011: 14) have also suggested that the small-scale Iron Age occupation at Thorney was possibly seasonally restricted, and perhaps related to 'those with flocks or herds at summer pasture'.

Romano-British

A steelyard weight and a sherd of possible greyware pottery were the only Romano-British artefacts recovered by the excavation. A small assemblage of Early Romano-British pottery was previously collected from the upper fills of the Middle Bronze Age enclosure ditches during the trial trenching (Peachey 2016a), suggesting that parts of the enclosure earthworks were still fractionally visible and open in the first century AD. It is highly likely that the Romano-British artefacts recovered from Swan's Nest relate to the settlement activity identified *c.*200 m to the south-west of the site at Land East of Brandon Road (Sleap and Savage 2016).

Significantly, the analysis of the environmental evidence from Swan's Nest has contextualised the Early–Late Romano-British remains at Land East of Brandon Road as likely representing the first instance of large-scale arable production to be carried out upon the clay-lands to the south-east of Swaffham, with a granary and multiple corn-drying ovens identified (*ibid.*). This ability to extensively cultivate an environment which countless prehistoric communities appear to have refrained from doing in the proceeding millennia likely relates to the significant technological advancements and social changes which occurred following the establishment of Roman administration in Britain across the latter half of the first century AD.

Medieval

Evidence for medieval activity at the site was only attested by a small collection of finds that had been residually and intrusively deposited in features dating to different periods, along with finds recovered from the topsoil. Medieval activity at Swan's Nest was most likely of a small-scale or transient nature, and it is highly likely that many of the artefacts found result from casual loss and discard. Equally, these finds may have been brought unintentionally to the site through night-soiling activities in the post-medieval period. It is highly likely that by this period, the ditches of the Middle Bronze Age enclosure were no longer open or visible.

As outlined in Chapter 1, following the cessation of activity at Brandon Road in the Late Romano-British period, significant occupation or agrarian exploitation of the Swan's Nest environs does not appear to recommence again until relatively recently, in the eighteenth–early nineteenth centuries. It can be speculated that this apparent reduction in activity after the Romano-British period could relate to the south-western retreat of the head of the River Gadder to Cockley Cley, with the area just over a kilometre south of Swan's Nest now a dry valley. In the absence of a natural abundant water source, which amongst other resources would have provided verdant seasonal pasturage, this heavy clay soil area may have become an increasingly unfavourable landscape in which to settle. It is also possible that both the pastoral and arable productivity of the clays had markedly diminished by the end of the Romano-British period, following centuries of extensive exploitation. However, this supposed shift in the local settlement pattern is unlikely to be the sole result of a precisely definable factor such as environmental change. It is highly likely instead that a plethora of social and economic changes were also factors in this.

Post-Medieval

Five ditches, a probable hedge-line and one pit identified during the excavation dated to the post-medieval period. The scarcity of medieval remains at the site may indicate that the area was not under heavy cultivation during the proceeding period, and that it was only through advancements in drainage techniques and artificial manuring technology during the eighteenth century (Sturgess 1966), that allowed for the exploitation of the heavy clay soils at Swan's Nest.

Cartographic and artefact evidence indicates that four of the ditches identified during the excavation (Ditch 1, Ditch 2, Ditch 3 and Ditch 4) represent agricultural field boundaries most likely instated at some point between the eighteenth and early nineteenth centuries. The way in which these ditches equally partitioned and arranged the landscape (mostly into 'strip fields' perpendicular to Watton Road), suggests they were likely instated during a single episode of land organisation.

The line of three of these ditches (Ditch 1, Ditch 3 and Ditch 4) correspond with field boundaries depicted on the 1840 Swaffham Tithe Map. The absence of Ditch 2 on this map indicates that by this point, this boundary had already been dissolved and the ditch backfilled. Ditch 3 and Ditch 4 were subsequently absent on the 1879–1886 First Edition Ordnance Survey Map, indicating that the original post-medieval field arrangements had become largely redundant by the late nineteenth century. Ditch 1 however remained open and in use up until the late twentieth century. The amalgamation of smaller fields into larger fields in this region during the later post-medieval period is a phenomenon identified at countless archaeological sites (e.g. Whitmore and Watkins 2016; White and Hodges 2019; Adams and Collyer 2020). This most likely relates to advancements in agricultural technology and associated economic factors during the nineteenth century, which favoured both larger fields and larger farms (Overton 1996).

The multifunctional nature of these ditches (beyond simply representing physical property boundaries) was demonstrated with Ditch 3, where a drainage pipe had been laid in the ditch prior to it being backfilled. This suggests that although there was a desire to dissolve this boundary, its drainage properties were acknowledged, and an attempt was made to preserve this function. The other sub-surface post-medieval remains at the site consisted of an infilled drainage ditch associated with the extant pond in Area A, an amorphous pit of unidentified function and a hedge-line associated with Ditch 3.

Through aerial photographic analysis undertaken as part of the archaeological investigations into Swan's Nest and its environs, several previously unknown features of probable post-medieval date were also identified surrounding the site. This included components related to agricultural field systems and an area of former coppice.

CHAPTER 7

Conclusion

The archaeological excavation at Swan's Nest has been significant in bringing to light past human activity dating from the Mesolithic to post-medieval periods. Regionally important Middle Bronze Age remains were identified, along with locally important Early Neolithic and earlier Iron Age activity. Aerial photographic analysis undertaken as part of this project has also identified nineteen previously unknown potential archaeological sites from the Swan's Nest environs, including significant possible prehistoric remains centred around TF 833 070.

The archaeological remains were well-preserved, with the prehistoric features being only minimally truncated by later historic and modern activity. Well-preserved ecofactual evidence was recovered (particularly the palynological and zooarchaeological remains), which allowed for informed interpretations of the sites previous environmental setting and possible use. A relatively small assemblage of finds was recovered from the prehistoric features; a characteristic typical of such sites in the region. The dating of these artefactual remains and the prehistoric features from which they were collected was corroborated by a strong sequence of radiocarbon dates.

The Middle Bronze Age remains centred around the discovery of an 'Ormesby-type' enclosure (after Gilmour et al. 2014), a monument of immense construction, which most likely had an important role in the cattle husbandry regimes of the communities that inhabited the Breckland clays between the fifteenth to twelfth centuries BC. In addition to the enclosure having had a practical purpose, probably in containing and defending herds of animals and acting as a place of congregation, it also likely had an important symbolic role within society; something possibly evidenced by the elaborate scale of the ditches and the burnt flint accumulations identified at the site.

The Swan's Nest enclosure has significant potential to help contribute towards the interpretation and further definition of the 'Ormesby-type' enclosure monument class; a task of regional importance in forwarding our collective understanding of Bronze Age societies and the establishment of settled agricultural communities in Norfolk. The Swan's Nest enclosure is notable in the way in which it appears to differ from other 'Ormesby-type' enclosures through its more regular form, the less defined predating/contemporary field system and that it was host to extensive 'burnt mound' activity.

Swan's Nest (along with a number of other sites recently discovered in the county), largely refutes the previously held assumption that in Norfolk there is no evidence for the 'expenditure of resources on monuments in the manner seen in Wessex' during the Bronze Age (Ashwin 1996: 54). Aerial photographic work carried out as part of this project has also illustrated that thorough analysis of cropmarks, even on clay soils, can be conducive for identifying enclosures and earthworks dating to this period; a process previously discussed as problematic in the county (see Ashwin 1996). Additionally, the evidence discovered posits that occupation of the boulder clays was not avoided in the Bronze Age and Iron Age, as has previously been suggested (see *ibid.*; Davies 1996), with the economies of these communities just more specialised.

Of greatest significance, Swan's Nest represents the most substantial discovery of prehistoric remains recorded from the Swaffham area to date, and consequently, these remains will be integral in shaping our understanding on the nature of the communities that once occupied the Breckland landscape throughout prehistory.

Appendices

APPENDIX A

Pollen: Table and Figures

Table 11. Total pollen, spores and non-tree pollen

Group	Pollen	Sample <44> Pit (597) deposit (598)	Sample <45> Pit (605) deposit (606)	Monolith <09> 28–29 cm Ditch (361) deposit (376)
A	*Pinus*	6		1
A	*Quercus*	3	1	1
A	*Alnus*	11	12	7
B	*Corylus*	1	2	1
B	*Salix*	2	1	
B	*Calluna*		8	
C	Poaceae	31	22	22
C	*Ranunculus-type*		1	
C	*Plantago major/media*		1	
C	*Plantago lanceolata*			1
C	*Rumex acetosella*	3		
C	*Compositae*			18
C	*Taraxacum-type*	21	27	37
C	*Circaea*	1		
E	*Pteridium*		3	8
E	*Polypodium*		1	
J	*Glomus*	15	6	
J	Undif. fungal spores/ *Sordariaceae*	6	2	
J	*Cercophora*	6		
F	Foraminfera			375
	Total land pollen	79	75	88

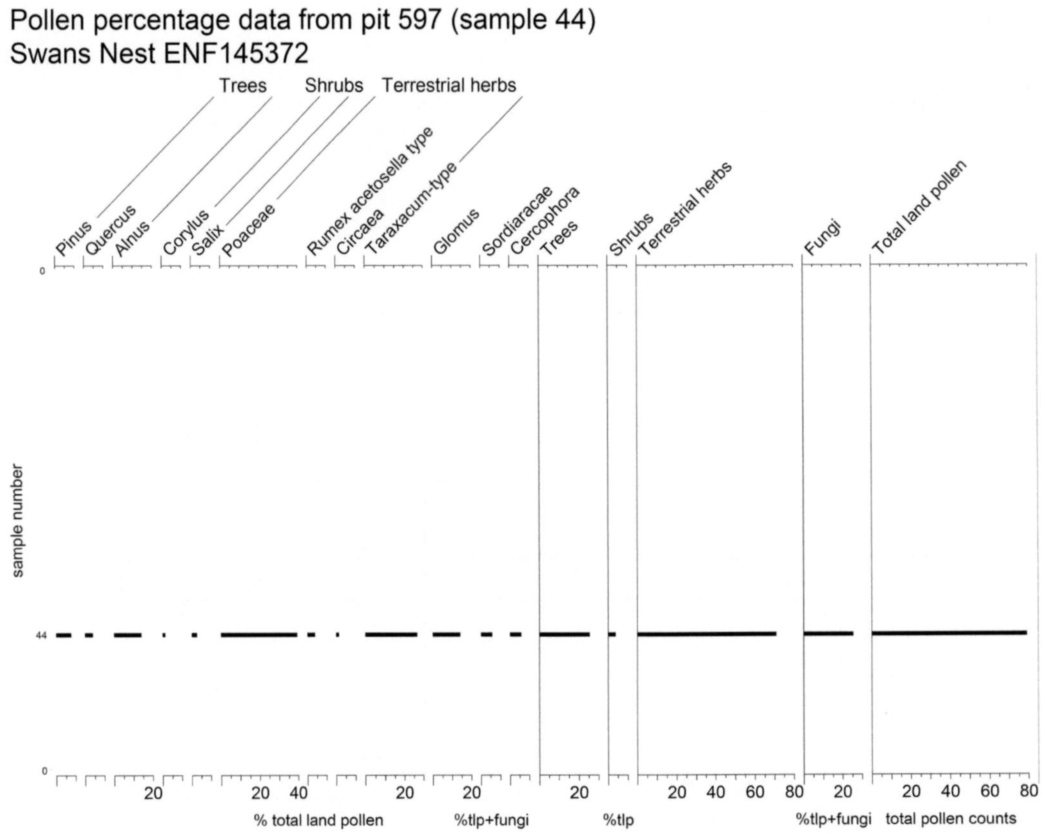

Figure 32. Pollen sample <44>. © c/o NPS Archaeology.

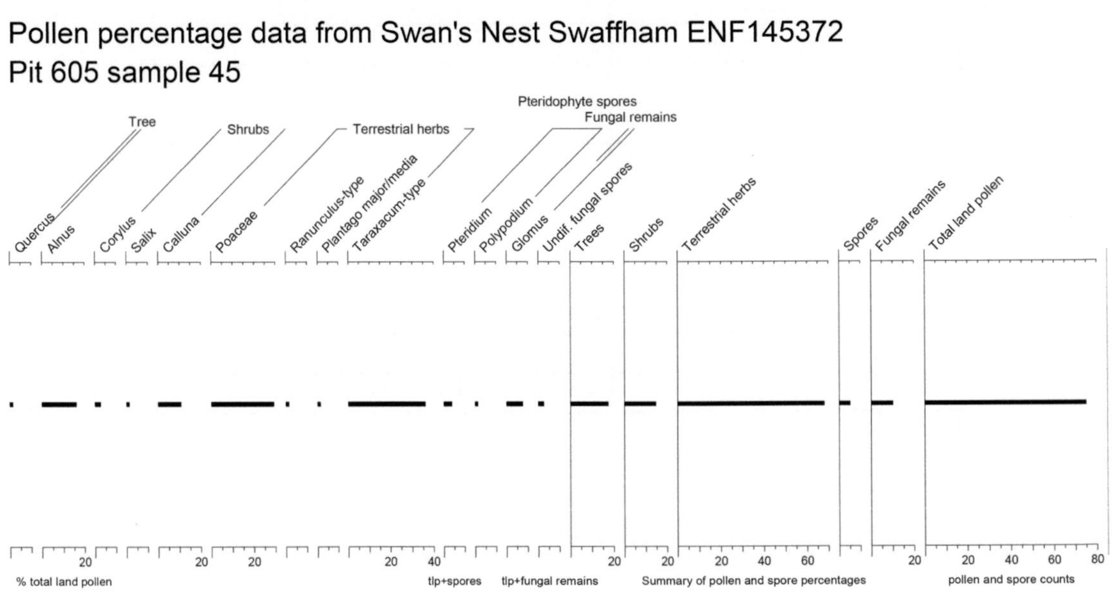

Figure 33. Pollen sample <45>. © c/o NPS Archaeology.

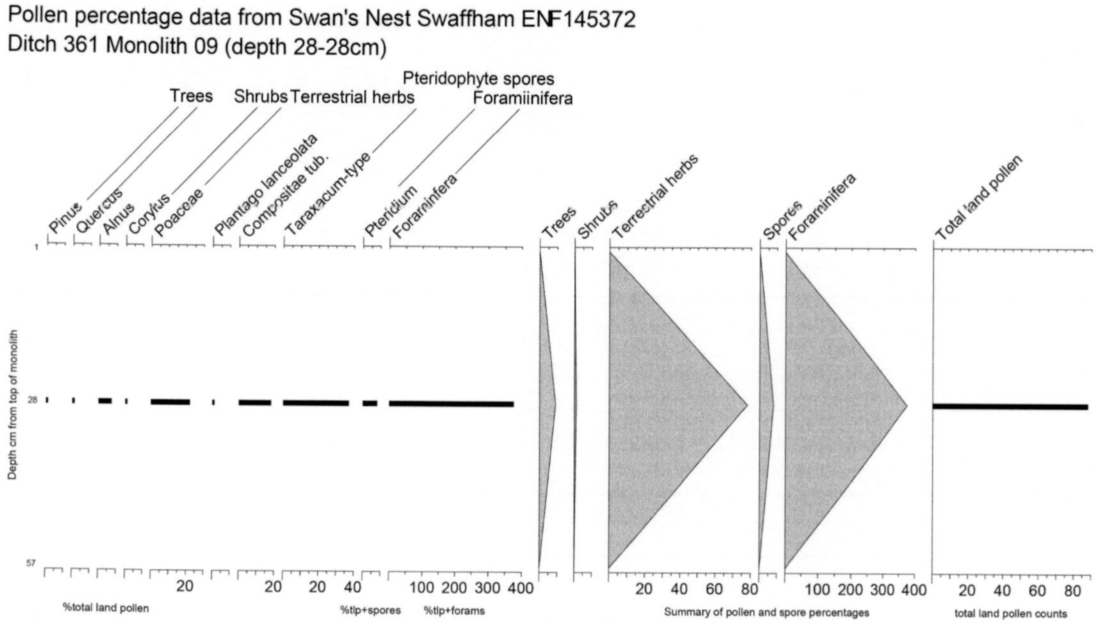

Figure 34. Pollen sample <09>. © c/o NPS Archaeology.

APPENDIX B

Micromorphology: Tables

Table 12. Type and percentage of post-depositional attributes within microstratigraphic units

Structure	Deposit type	Sample number	Context	MM Unit number	Weathering	Decay
					Translocation	Organic staining
					Iron- amorphous nodules	
MBA Enclosure Ditch Segment (361)	Fill	Monolith <9>	(376)	1	•••	••
				2	•••••	••

Key: ••••• Very abundant >20 per cent; •••• Abundant 10–20 per cent; ••• Many 5–10 per cent; •• Occasional 2–5 per cent; • Rare <2 per cent

Table 13. Description of sediment attributes for microstratigraphic units

Structure	Deposit type	Sample	Context number	MF type	Basal boundary	Particle size	Sorting	C: F ratio	Fine material	Groundmass	Colour	Related distribution	Microstructure	Inclusions: Orientation and distribution
MBA Enclosure Ditch Segment (361)	Fill	Monolith <9>	(376)	1	N/A	Sandy clay	Unsorted	40:60	Calcareous	Crysalltic. Dotted	Yellowish-brown/ greyish brown, PPL; Silvery/ golden (3rd order interference colours), XPL	Embedded	Spongey. Vughy.	Unoriented, unrelated. Random, unreferred.
				2	N/A	Sandy clay loam	Unsorted	60: 40	Mineral (Fe) and calcitic	Dotted. Crystallitic	Mid-brown, PPL; Orange brown/ greyish brown, XPL	Linked and coated, and embedded and coated	Spongey	Unoriented, unrelated. Random, unreferred.

Table 14. Percentage of inclusions within microstratigraphic units

Structure	Deposit type	Sample number	Context	MF Type	Thickness on slide (cm)	Bedding	Rock fragments		Minerals					Building materials & Sediment aggregates		Micro-artefacts	Bicarch	Organic/Plant remains		Micro-fossil
							Flint	Chalk	Calcite	Quartz	Microcline	Plagioclase	Muscovite	Fragments of Fe micropan	Silty clay sediment aggregate	Heated flint	Shell (fossil)	Dung (structure gone)	Charred wood	Spherulites
MBA Enclosure Ditch Segment 361	Fill	Monolith 9	376	1	N/A	Massive	***	***	**	=***	*							**		*
				2	N/A		***	****	**	=***	*	*		**	**	**	**		*	*

*Key: ****** Very dominant >70 per cent; ***** Dominant 50–70 per cent; **** Common 30–50 per cent; *** Frequent 15–30 per cent; ** Few 5–15 per cent; * Very few <5 per cent*

APPENDIX C

Archaeological Periods

Table 15. Archaeological periods. After English Heritage Periods list, recommended by Forum on Information Standards in Heritage, available at: <http://www.heritage-standards.org.uk/chronology/>

Period	Date from	Date to
Prehistoric	-1,000,000	43
Early Prehistoric	-1,000,000	-4,000
Palaeolithic	-1,000,000	-10,000
Lower Palaeolithic	-1,000,000	-150,000
Middle Palaeolithic	-150,000	-40,000
Upper Palaeolithic	-40,000	-10,000
Mesolithic	-10,000	-4,000
Early Mesolithic	-10,000	-7,000
Late Mesolithic	-7,000	-4,000
Late Prehistoric	-4,000	43
Neolithic	-4,000	-2,200
Early Neolithic	-4,000	-3,330
Middle Neolithic	-3,300	-2,900
Late Neolithic	-2,900	-2,200
Bronze Age	-2,600	-700
Early Bronze Age	-2,600	-1,600
Beaker	-2,300	-1,700
Middle Bronze Age	-1,600	-1,200
Late Bronze Age	-1,200	-700
Iron Age	-800	43
Early Iron Age	-800	-300
Middle Iron Age	-300	-100
Late Iron Age	-100	43
Roman	43	410
Post Roman	410	1900
Saxon	410	1065
Early Saxon	410	650
Middle Saxon	650	850
Late Saxon	850	1066
Medieval	1066	1540
Post-medieval	1540	1900
Modern	1900	2050
First World War	1914	1918
Second World War	1939	1945
Cold War	1945	1992
Unknown	–	–

Radiocarbon Dates

Charcoal fragments from contexts containing primarily hazel/alder and oak charcoal from both mature and twiggy specimens, along with a fragment of *Fabaceae* cotyledon and animal bone specimens from a large ungulate and a juvenile *Bos taurus* were selected for dating. The charcoal and nutshells were pre-treated using an acid-base-acid protocol. The bone collagen was extracted and pre-treated with alkali following the Longin Collagen Extraction Method. Pre-treated samples were then converted to graphite using methods described in *Introduction to Radiocarbon Determinations by the Accelerator Mass Spectrometry Method* (Beta Analytic *undated*). The samples were dated by Accelerator Mass Spectrometry at the Beta Analytic dating laboratory. The ages have been calculated using the reported delta-13C (δ13C) and delta-15N (δ15N) measured by Isotope Ratio Mass Spectrometer (IRMS) and calibrated using the maximum intercept method (Stuiver and Reimer 1986). The calibrated age ranges are quoted at 95.4 per cent confidence levels and with end points rounded outwards to ten years. They have been calibrated using BetaCal and the IntCal13 calibration dataset for the northern hemisphere (Reimer *et al.* 2013). The radiocarbon ages and stable isotope results are listed in Table 16.

Table 16. Swan's Nest radiocarbon and stable isotope measurements. Dates have been calibrated using IntCal13

Laboratory Number	Context	Sample Details	Radiocarbon Age (BP)	δ13C (‰)	δ15N (‰)	C:N (atomic)	Calibrated date (95% probability)
Beta-552728	(388)	Charcoal	3120±30	−24.9	–	–	1450–1290 cal. BC
Beta-552729	(376)	Charcoal	3140±30	−26.0	–	–	1500–1370 cal. BC (80.1%); 1350–1300 cal. BC (15.3%)
Beta-552730	(610)	*Fabaceae* cotyledon	4790±30	−23.9	-	–	3620–3520 cal. BC (78.2%); 3650–3610 cal. BC (17.2%)
Beta-552731	(106)	Large mammal bone collagen	3160±30	−21.3	+6.0	3.3	1510–1390 cal. BC (92.7%); 1340–1320 cal. BC (2.7%)
Beta-552732	(461)	*Bos taurus* bone collagen	3010±30	−21.8	+4.2	3.3	1310–1150 cal. BC (79%); 1390–1340 cal. BC (12%); 1150–1120 cal. BC (4.4%)

Bibliography

Adams, D. (2018). *Swan's Nest, Land West of Watton Road, Swaffham Norfolk. Written Scheme of Investigation – Archaeological Excavation*. Norwich: NPS Archaeology.

Adams, D., and Collyer, G. (2020). *Land at former Fakenham Academy, Highfield Road, Fakenham, Norfolk. Evaluation by Trial Trenching*. Norwich: NPS Archaeology, Report No. 2020/1056.

Albion Archaeology (2005). *Land off Brandon Road, Swaffham, Norfolk: Archaeological Field Evaluation*. Bedford: Albion Archaeology, Report No. 2005/62.

Allen, M. (2017). *Aller Court Farm, Aller, Somerset – Publication report for Proceedings of the Somerset Archaeological and Natural History Society*. Oxford: Oxford Archaeology.

Ames, J., and Sillwood, R. (2010). *An Archaeological Evaluation at Broadland Gate, Postwick, Norfolk*. Norwich: NAU Archaeology.

Anderson-Whymark, H. (2011). 'Neolithic to Early Bronze Age Pit Deposition Practices and the Temporality of Occupation in the Upper Thames Valley'. In H. Anderson-Whymark, and J. Thomas (eds), *Regional Perspectives on Neolithic Pit Deposition: Beyond the Mundane*. Oxford: Oxbow Books, 187–99.

Andrefsky, W. (2005). *Lithics: Macroscopic Approaches to Analysis*. 2nd edn. Cambridge: Cambridge University Press.

Andrew, R. (1984). *A Practical Pollen Guide to the British Flora. Technical Guide No.1*. Cambridge: Quaternary Research Association.

Ashwin, T. (1996). 'Neolithic and Bronze Age Norfolk', *Proceedings of the Prehistoric Society*, 62, 41–62.

Ashwin, T., and Flitcroft, M. (1999). 'The Launditch and Its Setting', *Norfolk Archaeology*, 43 (2), 217–57.

Austin, L. (1997). 'Palaeolithic and Mesolithic'. In J. Glazebrook (ed.), *Research and Archaeology: A Framework for the Eastern Counties, 1. Resource Assessment*. Dereham: East Anglian Archaeology, Occasional Papers, No. 3, 5–11.

Bagyaraj, D. J., and Varma, A. (1995). 'Interaction between Arbuscular Mycorrhizal Fungi and Plants: Their Importance in Sustainable Agriculture in Arid and Semi-arid Tropics'. In J. G. Jones (ed.), *Advances in Microbial Ecology*, vol. 14. Oxford: Blackwell Scientific Publications, 119–36.

Baker, P., and Worley, F. (2014). *Animal Bones and Archaeology – Guidelines for Best Practice*. London: English Heritage.

Banerjea, R. Y., Badura, M., Brown, A., Morandi, L. F., Marcinkowski, M., Valk, H., Ismail-Meyer, K., and Pluskowski, A. (2020). 'Feeding the Crusades: Archaeobotany, Animal Husbandry and Livestock Alimentation on the Baltic Frontier', *Environmental Archaeology: The Journal of Human Palaeoecology*, 25 (2), 135–50.

Barfield, L. H., and Hodder, M. (1987). 'Burnt Mounds as Saunas, and the Prehistory of Bathing', *Antiquity*, 61, 370–9.

Bates, S. J. (2008). *Archaeological Works on the Line of the Bacton to Great Yarmouth Gas Pipeline Norfolk 1997–9*. Norwich: NPS Archaeology.

Bates, S. J., and Wiltshire, P. E. J. (2000). 'Excavation of a Burnt Mound at Feltwell Anchor, Norfolk, 1992', *Norfolk Archaeology*, 43, 389–414.

Beadsmore, E. (2006). 'Earlier Neolithic Flint'. In D. Garrow, S. Lucy, and D. Gibson (eds), *Excavations at Kilverstone, Norfolk: An Episodic Landscape History*. Dereham: East Anglian Archaeology, No. 113, 53–70.

Behre, K. E. (1988). 'The Rôle of Man in European Vegetation History'. In B. Huntley, and T. Webb (eds), *Vegetation History. Handbook of Vegetation Science*, vol. 7. Dordrecht: Springer Kluwer Academic Publishers, 633–72.

Beta Analytic (*undated*). *Introduction to Radiocarbon Determinations by the Accelerator Mass Spectrometry Method*. Miami, FL: Beta Analytic.

Bishop, B. (2011). 'Earlier Neolithic Flintwork'. In B. Bishop, and J. Proctor (eds), *Settlement, Ceremony and Industry on Mousehold Heath: Excavations at Laurel Farm (Phase II), Broadland Business Park, Thorpe St. Andrew, Norfolk*. London: Pre-Construct Archaeology, PCA Monograph, No. 13, 31–48.

Bishop, B. (2018). 'The Lithic Material'. In D. Jacques, T. Phillips, and T. Lyons (eds), *Blick Mead: Exploring the 'First Place' in the Stonehenge Landscape*. Oxford: Peter Lang, 67–126.

Blackbourn, K. (2021). *A Middle Bronze Age Enclosure and Post-Medieval Activity on Land West of Brandon Road, Swaffham, Norfolk. Archaeological Excavation Report*. Cambridge: Oxford Archaeology, Report No. 2502.

Branch, N., Canti, M., Clark, P., and Turney C. (2005). *Environmental Archaeology Theoretical and Practical Approaches*. London: Hodder Arnold.

Breckland Society (2021). *The Breckland Society*. <http://www.brecsoc.org.uk/>, accessed 30 Feburary 2021.

British Geological Survey (2018). *Geology of Britain Viewer*. <http://mapapps.bgs.ac.uk/geologyofbritain/home.html>, accessed 10 September 2020.

Brudenell, M. (2012). *Pots, Practice and Society: An Investigation of Pattern and Variability in the Post-Deverel Rimbury Ceramic Tradition of East Anglia*. University of York, Unpublished PhD Thesis.

Bull, K., and Smith, L. (2016). *Swan's Nest, Brandon Road, Swaffham, Norfolk. An Archaeological Evaluation (Trial Trenching)*. Bury St Edmunds: Archaeological Solutions Ltd, Report No. 5113.

Bullock, P., Fedoroff, N., Jongerius, A., Stoops, G., and Tursina, T. (1985). *Handbook for Thin Section Description*. Wolverhampton: Waine Research.

Canti, M. G. (2003). 'Earthworm Activity and Archaeological Stratigraphy: A Review of Products and Processes', *Journal of Archaeological Science*, 30, 135–48.

Canti, M. G. (2007). 'Deposition and Taphonomy of Earthworm Granules in Relation to their Interpretive Potential in Quaternary Stratigraphy', *Journal of Quaternary Science*, 22 (2), 111–18.

Canti, M. G., and Brochier, J. E. (2017). 'Faecal Spherulites'. In C. Nicosia, and G. Stoops (eds), *Archaeological Soil and Sediment Micromorphology*. Chichester: John Wiley & Sons Ltd, 51–4.

Chadwick, A. M. (2012). 'Routine Magic, Mundane Ritual: Towards a Unified Notion of Depositional Practice', *Oxford Journal of Archaeology*, 31 (3), 283–315.

Champion, T. (2011). 'Later Prehistory'. In P. Booth, T. Champion, S. Foreman, P. Garwood, H. Glass, J. Munby, and A. Reynolds (eds), *On Track: The Archaeology of High Speed 1 Section 1 in Kent*. Oxford: Oxford Wessex Archaeology & High Speed.

Chapman, J. (2000). 'Pit Digging and Structured Deposition in the Neolithic and Copper Age', *Proceedings of the Prehistoric Society*, 66, 61–87.

Clark, J. G. D., and Fell, C. I. (1953). 'The Early Iron Age Site of Micklemoor Hill, West Harling, Norfolk, and It's Pottery', *Proceedings of the Prehistoric Society*, 19 (1), 1–40.

Clark, J., Higgs, E. S., and Longworth, I. H. (1960). 'Excavations at the Neolithic Site at Hurst Fen, Mildenhall, Suffolk (1954, 1957 and 1958)', *Proceedings of the Prehistoric Society*, 26, 202–45.

Cornwall, I. W. (1974). *Bones for the Archaeologist*. Worthing: Littlehampton Book Services Ltd.

Courty, M. A., Goldberg, P., and Macphail, R. (1989). *Soils and Micromorphology in Archaeology*. Cambridge: Cambridge University Press.

Couzens, D., Swash, A., Still, R., and Dunn, J. (2017). *Britain's Mammals. A Field Guide to the Mammals of Britain and Ireland*. Woodstock: Princeton University Press.

Cox, C. (2013). *Interpretation of Aerial Photographs for Archaeology in Hertfordshire. 2010–2013 HER Update*. Swindon: Air Photo Services, Report No. CC0209 03 01.

Cox, C. (2017). *Norfolk Vanguard Offshore Wind Farm (Terrestrial Archaeology). Heritage Assessment of Aerial Imagery*. Swindon: Air Photo Services, Report No. 217 04 01.

Crawley, P. E. (2014). *Archaeological Strip, Map and Sample Excavation at the Postwick Hub, Norwich: Assessment Report and Updated Project Design*. Norwich: NPS Archaeology.

Crowson, A. (2004). *Hot Rocks in the Norfolk Fens: The Excavation of a Burnt Flint Mound at Northwold, 1994–5*. Dereham: East Anglian Archaeology, Occasional Papers, No. 16.

Cunliffe, B. (2003). *Danebury Hillfort*. Stroud: Tempus Publishing Ltd.

Curl, J. (2020a). 'Faunal Remains'. In J. C. White (ed.), *Swan's Nest, Land West of Watton Road, Swaffham, Norfolk – Archaeological Excavation of Early Neolithic Occupation, a Middle Bronze Age Enclosure and Earlier Iron Age Activity*. Norwich: NPS Archaeology, Report No. 2020/1011, 146–52.

Curl, J. (2020b). 'Molluscs'. In J. C. White (ed.), *Swan's Nest, Land West of Watton Road, Swaffham, Norfolk – Archaeological Excavation of Early Neolithic Occupation, a Middle Bronze Age Enclosure and Earlier Iron Age Activity*. Norwich: NPS Archaeology, Report No. 2020/1011, 153–56.

Darvill, T. (1996). 'Neolithic Buildings in England, Wales and the Isle of Mann'. In T. Darvill, and J. Thomas (eds), *Neolithic Houses in Northwest Europe and Beyond*. Oxford: Oxbow Books, 77–112.

Darvill, T. (2010). *Prehistoric Britain*. 2nd edn. Abingdon: Routledge.

Davies, J. A. (1996). 'Where Eagles Dare: The Iron Age of Norfolk', *Proceedings of the Prehistoric Society*, 62, 63–92.

Davies, J. A. (1999). 'Patterns, Power and Political Progress in Iron Age Norfolk'. In J. Davies, and T. Williamson (eds), *Land of the Iceni – The Iron Age in Northern East Anglia*. Norwich: Centre of East Anglian Studies, 14–44.

Davis, S. (1992). *A Rapid Method for Recording Information about Mammal Bones from Archaeological Sites*. London: English Heritage AML, 71/92.

Evans, J. G. (1972). *Land Snails in Archaeology*. Cambridge: Academic Council Press.

Fitter, R., Fitter, A., and Blamey, M. (1985). *The Wild Flowers of Britain and Northern Europe*. 4th edn. London: Collins.

Fryer, V. (2019). *An Assessment of the Charred Plant Macrofossils and Other Remains from Over, Cambridgeshire (OVE18(2))*. Assessment report for Cambridge Archaeological Unit.

Garrow, D., Beadsmoore, E., and Knight, M. (2005). 'Pit Clusters and the Temporality of Occupation: An Earlier Neolithic Site at Kilverstone, Norfolk', *Proceedings of the Prehistoric Society*, 71, 139–57.

Garrow, D., Lucy, S., and Gibson, D. (2006). *Excavations at Kilverstone, Norfolk: Neolithic Pits, Later Prehistoric, Romano British and Anglo-Saxon Occupation and Later Activity*. Dereham: East Anglian Archaeology, No. 113.

Gilmour, N., Horlock, S., Mortimer, R., and Tremlett, S. (2014). 'Middle Bronze Age Enclosures in the Norfolk Broads: A Case Study at Ormesby St Michael, England', *Proceedings of the Prehistoric Society*, 80, 141–57.

Gilmour, N., and Mortimer, R. (2012). *Early Prehistoric Settlement, Field Systems and Enclosure at Ormesby St Michael, Norfolk*. Oxford: Oxford Archaeology, Report No. 1170.

Grant, A. (1984). 'Animal Husbandry'. In B. Cunliffe (ed.), *Danebury: An Iron Age Hillfort in Hampshire Vol 2 The Excavations 1969–1978: The Finds*. London: Council for British Archaeology, 496–548.

Grant, A. (1989). 'Animals and Ritual in Early Britain: The Visible and the Invisible', *Anthropozoologica*, troisième numéro special, 79–86.

Green, F. (2008). *An Archaeological Excavation at Three Score Community Residential Development, Bowthorpe, Norfolk*. Norwich: NPS Archaeology, Report No. 1076.

Green, M. (2018). *Marham Park, Bury St Edmunds, Suffolk. Post Excavation Assessment Report*. Needham Market: Suffolk Archaeology CIC, Report No. 2018/040.

Gregory, T. (1992). *Excavations in Thetford, 1980–1982, Fison Way*. Dereham: East Anglian Archaeology, No. 53.

Grimm, E. (1993). *TILIA: A Pollen Program for Analysis and Display*. Springfield, IL: Illinois State Museum.

Grimm, E. (2011). 'Environmental Evidence'. In K. Egging Dinwiddy, and P. Bradley (eds), *Prehistoric Activity and a Romano-British Settlement at Poundbury Farm, Dorchester, Dorset*. Salisbury: Wessex Archaeology, 133–42.

Hamilton, W. D., Haselgrove, C., and Gosden, C. (2015). 'The Impact of Bayesian Chronologies on the British Iron Age', *World Archaeology*, 47 (4), 642–60.

Hather, J. G. (2000). *The Identification of the Northern European Woods*. London: Archetype Publications.

Healy, F. (1988). *The Anglo-Saxon Cemetery at Spong Hill, North Elmham, Part VI: Occupation during the Seventh to Second Millennium BC*. Dereham: East Anglian Archaeology, No. 39.

Hickling, S. (2017). *Brief for Archaeological Excavation at Swan's Nest, Brandon Road, Swaffham, Norfolk*. Gressenhall: Norfolk County Council Environmental Services.

Hill, J. D. (1996). 'The Identification of Ritual Deposits of Animals. A general perspective from a specific study of "special animal deposits" from the Southern English Iron Age'. In S. Anderson and K. V. Boyle (eds), *Ritual Treatment of Human and Animal Remains: Proceedings of the First Meeting of the Osteoarchaeological Research Group Held in Cambridge on 8th October 1994*. Oxford: Oxbow Books, 17–32.

Hill, J. D. (1999). 'Settlement, Landscape and Regionality: Norfolk and Suffolk in the Pre-Roman Iron Age of Britain and Beyond'. In J. Davies, and T. Williamson (eds), *Land of the Iceni – The Iron Age in Northern East Anglia*. Norwich: Centre of East Anglian Studies, 185–207.

Hills, C., and Lucy, S. (2013). *Spong Hill Part IX: Chronology and Synthesis*. Cambridge: McDonald Institute for Archaeological Research.

Hills, C., and Wade-Martins, P. (1976). *The Anglo-Saxon Cemetery at the Paddocks, Swaffham. Norfolk*. Dereham: East Anglian Archaeology, No. 2.

Hobbs, B. (2002). *Report on an Archaeological Evaluation at 87–95 Market Place, Swaffham, Norfolk*. Norwich: Norfolk Archaeological Unit, Report No. 666.

Hobbs, B. (2004). *An Archaeological Excavation at the Paddocks, Cley Road, Swaffham, Norfolk*. Norwich: Norfolk Archaeological Unit, Report No. 937.

Hudson, J. D., and Palmer, T. J. (1976). 'A Euryhaline Oyster from the Middle Jurassic and the Origin of the True Oyster', *Palaeontology*, 19 (1), 79–93.

Hutton, J. (2008). *Excavations at Langtoft, Lincolnshire. The Glebe Land*. Cambridge: Cambridge Archaeological Unit, Report No. 837.

Jacobi, R. (1984). 'The Mesolithic of Northern East Anglia and Contemporary Territories'. In C. Barringer (ed.), *Aspects of East Anglian Pre-History: 20 Years after Rainbird Clarke*. Norwich: Geo Books, 43–76.

Jeffery, P. (1991). 'Burnt Mounds, Fulling and Early Textiles'. In M. A. Hodder, and L. H. Barfield (eds), *Burnt Mounds and Hot Stone Technology*. West Bromwich: Sandwell Metropolitan Borough Council, 97–107.

Kooistra, M. J., and Pulleman, M. M. (2018). 'Features Related to Faunal Activity'. In G. Stoops, V. Marcelino, and F. Mees (eds), *Interpretation of Micromorphological Features of Soils and Regoliths*. Amsterdam: Elsevier, 447–70.

Last, J. (forthcoming). 'The Neolithic Period'. In *East Anglian Regional Framework*. East of England: ALGAO.

Lawson, A. (1983). *The Archaeology of Witton Near North Walsham, Norfolk*. Barnsley: East Anglian Archaeology, No. 18.

Longworth, I., Ellison, A., and Rigby, V. (1988). *Excavations at Grimes Graves, Norfolk 1972–1976. Fascicle 2. The Neolithic, Bronze Age and Later Pottery*. London: British Museum Publications Limited.

Lord, J. (1993). *The Nature and Subsequent Uses of Flint*. Privately Published.

Mackenzie, W. S., and Adams, A. E. (1994). *A Colour Atlas of Rocks and Minerals in Thin Section*. New York: John Wiley and Sons.

Mackenzie, W. S., and Guilford, C. (1980). *Atlas of Rock Forming Minerals*. Harlow: Longman Group Ltd.

Macphail, R. I. (1994). 'The Reworking of Urban Stratigraphy by Human and Natural Processes'. In H. K. Kenward (ed.), *Urban-Rural Connexions: Perspectives from Environmental Archaeology*. Barnsley: Oxbow, 13–43.

Marsh, B. P. (2011). *Land off Brandon Road and Watton Road, Swaffham, Norfolk. An Archaeological Evaluation*. Upton upon Severn: Stratascan.

Moffett, L., Robinson, M. A., and Straker, V. (1989). 'Cereals, Fruit and Nuts: Charred Plant Remains from Neolithic Sites in England and Wales and the Neolithic Economy'. In A. Milles, D. Williams, and N. Gardner (eds), *The Beginnings of Agriculture*. Oxford: B.A.R. International Series, No. 496, 243–61.

Moore, P. D., Webb, J. A., and Collinson, M. E. (1991). *Pollen Analysis*. 2nd edn. Oxford: Blackwell Scientific Publications.

Morris, J. (1984). *Domesday Book, 33 Norfolk, Part I and Part II*. Chichester: Phillimore and Co.

Munsterman, D., and Kerstholt, S. (1996). 'Sodium Polytungstate, a New Non-Toxic Alternative to Bromoform in Heavy Liquid Separation', *Review of Palaeobotany and Palynology*, 91 (1), 417–22.

Murphy, P. (1984). 'Prehistoric Environments and Economies'. In C. Barringer (ed.), *Aspects of East Anglian Pre-History: 20 Years after Rainbird Clarke*. Norwich: Geo Books, 13–30.

Murphy, P. (2000). 'Environmental and Botanical Evidence'. In T. Ashwin, and S. Bates (eds), *Excavations on the Norwich Southern Bypass, 1989–91 Part 1: Excavations at Bixley, Caistor St Edmund, Trowse, Cringleford and Little Melton*. Barnsley: East Anglian Archaeology, No. 91, 217–23.

Nakagawa, T., Brugiapaglia, E., Digerfeldt, G., Reille, M., Beaulieu, J-L De., and Yasuda, Y. (1998). 'Dense-Media Separation as a More Efficient Pollen Extraction Method for Use with Organic Sediment/Deposit Samples: Comparison with the Conventional Method', *Boreas*, 27, 15–24.

Natural History Museum (2001). *British Mesozoic Fossils*. 6th edn. London: Natural History Museum.

Ó Drisceoil, D. A. (1988). 'Burnt Mounds: Cooking or Bathing?', *Antiquity*, 62, 671–80.

Overton, M. (1996). 'Re-establishing the English Agricultural Revolution', *Agricultural History Review*, 44 (2), 1–20.

Peachey, A. (2016a). 'Late Iron Age and Roman Pottery'. In K. Bull and L. Smith (eds), *Swan's Nest, Brandon Road, Swaffham, Norfolk. An Archaeological Evaluation (Trial Trenching)*. Bury St. Edmunds: Archaeological Solutions Ltd, Report No. 5113, 29–31.

Peachey, A. (2016b). 'Fired Clay and Ceramic Building Materials'. In K. Bull and L. Smith (eds), *Swan's Nest, Brandon Road, Swaffham, Norfolk. An Archaeological Evaluation (Trial Trenching)*. Bury St. Edmunds: Archaeological Solutions Ltd, Report No. 5113, 29–31.

Percival, S. (2018). *Prehistoric Pottery from the Norwich Northern Distributor Road*. Report for Oxford Archaeology East.

Phillips, T., and Mortimer, R. (2012). *Clay Farm, Trumpington, Cambridgeshire: Post-excavation Assessment and Updated Project Design*. Oxford: Oxford Archaeology East, Report No. 1294.

Pickstone, A., and Mortimer, R. (2011). *The Archaeology of Brigg's Farm, Prior's Fen, Thorney, Peterborough*. Oxford: Oxford Archaeology, Report No. 1094.

Pollard, J. (2001). 'The Aesthetics of Depositional Practice', *World Archaeology*, 33, 315–35.

Prehistoric Ceramics Research Group (2010). *The Study of Prehistoric Pottery: General Policies and Guidelines for Analysis and Publication*. Prehistoric Ceramics Research Group, Occasional Papers, No. 1 and No. 2.

Pryor, F. (2001). *The Flag Fen Basin Archaeology and Environment of a Fenland Landscape*. London: English Heritage.

Quinn, S., and Peachey, A. (2012). *Swan's Nest, Brandon Road, Swaffham, Norfolk. An Archaeological Evaluation (Field Walking and Metal Detecting Survey)*. Bury St Edmunds: Archaeological Solutions Ltd, Report No. 4025.

Reimer, P. J., Bard, E., Bayliss, A., Beck, W., Blackwell, P. G., Ramsey, C. B., Buck, C. E., Cheng, H., Edwards, R. L., Friedrich, M., Grootes, P. M., Guilderson, T. P., Haflidason, H., Hajdas, I., Hatté, C., Heaton, T. J., Hoffmann, D. L., Hogg, A. G., Hughen, K. A., Kaiser, K. F., Kromer, B., Manning, S. W., Niu, M., Reimer, R. W., Richards, D. A., Scott, M., Southon, J. R., Staff, R. A., Turney, C. S. M., and van der Plicht, J. (2016). 'IntCal13 and Marine13 Radiocarbon Age Calibration Curves 0–50,000 Years cal BP', *Radiocarbon*, 55 (4), 1869–1887.

Rentzel, P., Nicosia, C., Gebhardt, A., and Brönnimann, D. (2017). 'Trampling, Poaching and the Effect of Traffic'. In C. Nicosia, and G. Stoops (eds), *Archaeological Soil and Sediment Micromorphology*. Chichester: John Wiley & Sons Ltd, 281–98.

Schweingruber, F. H. (1990). *Microscopic Wood Anatomy*. 3rd edn. Birmensdorf: Swiss Federal Institute for Forest, Snow and Landscape Research.

Scollar, I. (2002). 'Making Things Look Vertical'. In R. H. Bewley, and W. Raczkowski (eds), *Aerial Archaeology: Developing Future Practice*. Ohmsha: Ios Press, NATO Science Series, 337, 166–72.

Scollar, I., and Palmer, R. (2008). 'Using Google Earth Imagery', *AARG News*, 37, 15–21.

Simpson, I. (1998). 'Early Land Management at Tofts Ness, Sanday, Orkney: The Evidence of Thin Section Micromorphology'. In C. M. Mills, and G. Coles (eds), *Life on the Edge: Human Settlement and Marginality*. Barnsley: Oxbow, 91–8.

Sleap, J., and Savage, R. D. (2016). *Phase 3, Land East of Brandon Road, Swaffham, Norfolk. Archaeological Scheme of Strip, Map and Record*. Lincoln: Pre-construct Archaeological Services Ltd, Report No. 1427.

Smith, I. F. (1965). *Windmill Hill and Avebury: Excavations by Alexander Kieller 1925–39*. Oxford: Clarendon Press.

Smith, L. (2012). *Swan's Nest, Brandon Road, Swaffham, Norfolk. An Archaeological Evaluation (Trial Trenching)*. Bury St Edmunds: Archaeological Solutions Ltd, Report No. 4142.

Stace, C. (2010). *New Flora of the British Isles*. 3rd edn. Cambridge: Cambridge University Press.

Stephens, N. (1990). *Natural Landscapes of Britain from the Air*. Cambridge: Cambridge University Press.

Stockmarr, J. (1971). 'Tablets with Spores Used in Absolute Pollen Analysis', *Pollen et Spores*, 13, 615–21.

Stoops, G. (2003). *Guidelines for Analysis and Description of Soil Thin Sections*. Madison: Soil Science Society of America.

Stoops, G., Marcelino, V., and Mees, F. (2018). *Interpretation of Micromorphological Features of Soils and Regoliths*. Amsterdam: Elsevier.

Stuiver, M. and Reimer, P. J. (1986). 'A computer program for radiocarbon age calibration'. In M. Stuiver, and R. S. Kra (eds), Proceedings of the 12th International C14 Conference, *Radiocarbon*, 28 (2B), 1022–1030.

Sturgess, R. W. (1966). 'The Agricultural Revolution on the English Clays', *Agricultural History Review*, 14 (2), 104–21.

Summers, J. (2016). 'The Environmental Samples'. In K. Bull and L. Smith (eds), *Swan's Nest, Brandon Road, Swaffham, Norfolk. An Archaeological Evaluation (Trial Trenching)*. Bury St. Edmunds: Archaeological Solutions Ltd, Report No. 5113, 33–9.

Thomas, J. (1999). *Understanding the Neolithic*. London: Routledge.

Tröels-Smith, J. (1955). 'Characterisation of Unconsolidated Sediments', *Danmarks Geologiske Undersolgelse Series*, 4 (3), 38–73.

Van Hoeve, M. L., and Hendrikse, M. (1998). *A Study of Non-Pollen Objects in Pollen Slides: The Types as Described by Dr Bas van Geel and Colleagues – Unpublished Compilation*. Utrecht: University of Utrecht.

Vepraskas, M. J., Lindbo, D. L., and Stolt, M. H. (2018). 'Redoximorphic Features'. In G. Stoops, V. Marcelino, and F. Mees (eds), *Interpretation of Micromorphological Features of Soils and Regoliths*. Amsterdam: Elsevier, 425–46.

Von den Driesch, A. (1976). *A Guide to the Measurements of Animal Bones from Archaeological Sites*. Cambridge, MA: Peabody Museum, Bulletin 1.

Wainwright, G. (1972). 'The Excavation of a Neolithic Settlement on Broome Heath, Ditchingham, Norfolk', *Proceedings of the Prehistoric Society*, 38, 1–97.

White, J. C. (2019). *Swan's Nest, Land West of Watton Road, Swaffham Norfolk – Archaeological Excavation: Assessment Report and Updated Project Design*. Norwich: NPS Archaeology, Report No. 2019/1011.

White, J. C. (2020). *Swan's Nest, Land West of Watton Road, Swaffham, Norfolk – Archaeological Excavation of Early Neolithic Occupation, a Middle Bronze Age Enclosure and Earlier Iron Age Activity*. Norwich: NPS Archaeology, Report No. 2020/1011.

White, J. C., and Hodges, L. (2019). *Land East of Artiss Close and Rotheram Road, Bildeston, Suffolk. Report on Informative Trial Trenching*. Norwich: NPS Archaeology, Report No. 2019/101055.

Whitmore, D. and Watkins, P. (2016). *Watlington Quarry, Watlington and Tottenhill, Norfolk – Archaeological Excavation Assessment and Updated Project Design*. Norwich: NPS Archaeology, Report No. 2016/1038.
Whittaker, J. (1994). *Flintknapping: Making and Understanding Stone Tools*. Austin: University of Texas Press.
Whittle, A., Healy, F., and Bayliss A. (2011). *Gathering Time. Dating the Early Neolithic Enclosures of Southern Britain and Ireland*. Barnsley: Oxbow.
Wymer, J. (1977). *Gazetteer of Mesolithic Sites in England and Wales*. London: CBA Research Report, No. 22.
Wymer, J., and Robins, P. (1995). 'A Mesolithic Site at Great Melton', *Norfolk Archaeology*, 42, 125–47.
Yates, D. (2007). *Land, Power and Prestige, Bronze Age Field Systems in Southern England*. Oxford: Oxbow.

Index

Studies in European Archaeology

Series Editors

Joshua C White, David Jacques and Graeme Davis

The aim of the series is to publish academically and theoretically stimulating reports across different areas of archaeology in Europe from the prehistoric to early modern worlds. The series serves as a platform for original contributions through excavation results, fieldwork and datasets, and interpretation of archaeological topics. It aims to support the archaeological community both by disseminating findings and serving as a forum for comparative enquiry.

Volume 1

Joshua C White

Swan's Nest, Swaffham, Norfolk: Excavating the Prehistory of the Breckland Clays

2022. ISBN: 978-1-80079-104-6